Contents

HOW TO USE THIS BOOK

The Little Book of Mathematical Principles, Theories, & Things is an easy-to-use comprehensive guide to mathematics. It features over 130 entries on key principles or theories essential to understanding the subject. Written in an easily accessible manner, the Little Book explains sometimes very difficult concepts and theories, putting them in their historical context, giving background information on the experts who proposed them in the first place, analysing influences and proposing, where relevant, links to other related entries. The book also features tables, equations and illustrations, and ends with a glossary, and a comprehensive index.

The Little Book of Mathematical Principles, Theories, & Things is arranged chronologically and the country of origin is listed, where appropriate. Each entry includes a clear main heading, the person or people responsible for the discovery, birth and death dates, where relevant, followed by a short introductory paragraph explaining the concept concisely. In some cases, the main essay is also cross referenced to linked subjects. The key on the opposite page explains the order of information in each entry.

THE
LITTLE
BOOK *OF*

MATHEMATICAL
PRINCIPLES,
THEORIES, & THINGS

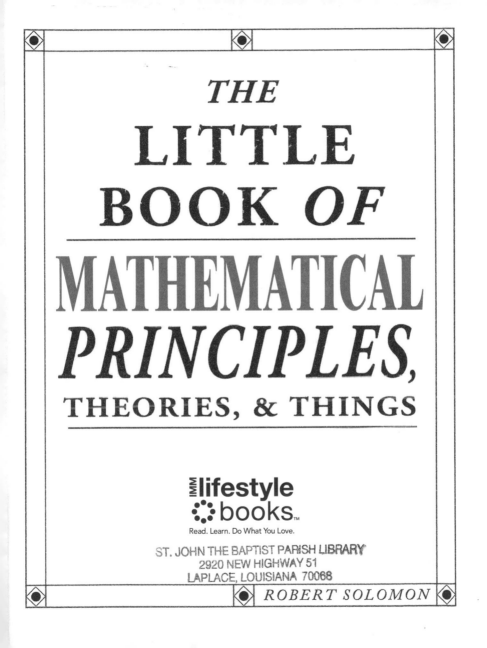

ᴵᴹᴹ lifestyle
:: books.
Read. Learn. Do What You Love.

ROBERT SOLOMON

Published 2016 — IMM Lifestyle Books
www.IMMLifestyleBooks.com

IMM Lifestyle Books are distributed
in the United Kingdom by Grantham Book Service,
Trent Road, Grantham, Lincolnshire, NG31 7XQ.

Text © 2016 by Dr Bob Solomon
Illustrations © 2016 by IMM Lifestyle Books

ISBN 978–1–5048–0053–2

Manufactured in China

10 9 8 7 6 5 4 3 2 1

The name of the mathematical principle.

The year or years of discovery, followed by the country or countries of discovery.

The name or names of the person or persons responsible for the principle, followed by their birth and death dates (where relevant).

A brief summary of the principle.

A concise one- to two-page entry, explaining the mathematical principle's importance and putting it in context.

Some entries are supported by explanatory diagrams, tables or graphs.

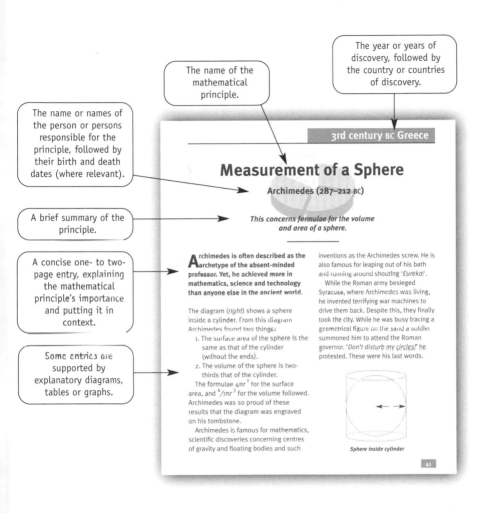

Measurement of a Sphere

Archimedes (287–212 BC)

This concerns formulae for the volume and area of a sphere.

Archimedes is often described as the archetype of the absent-minded professor. Yet, he achieved more in mathematics, science and technology than anyone else in the ancient world.

The diagram (*right*) shows a sphere inside a cylinder. From this diagram Archimedes found two things:
1. The surface area of the sphere is the same as that of the cylinder (without the ends).
2. The volume of the sphere is two-thirds of that of the cylinder.

The formulae $4\pi r^2$ for the surface area, and $^4/_3\pi r^3$ for the volume followed. Archimedes was so proud of these results that the diagram was engraved on his tombstone.

Archimedes is famous for mathematics, scientific discoveries concerning centres of gravity and floating bodies and such inventions as the Archimedes screw. He is also famous for leaping out of his bath and running around shouting '*Eureka!*'.

While the Roman army besieged Syracuse, where Archimedes was living, he invented terrifying war machines to drive them back. Despite this, they finally took the city. While he was busy tracing a geometrical figure on the sand a soldier summoned him to attend the Roman governor. '*Don't disturb my circles!*' he protested. These were his last words.

Sphere inside cylinder

41

Writing Numbers

The place value system uses only a finite number of symbols to write any number.

In some number systems, there are different symbols for each power of 10. In a place value system, only a small number of symbols are used.

In the Ancient Egyptian number system, dating from about 3000 BC, there were symbols for units, symbols for 10s, and so on. The number 365 was written:

℮℮℮ ∩∩∩ ||
∩∩∩ |||

where | represents a unit, ∩ represents 10 and ℮ represents 100.

The Chinese system writes numbers much as we say them. We say "three hundred and sixty-five:" in other words, so many hundreds, so many tens, and so many units. The number 365 is written as shown below.

三	百	六	十	五
3	100	6	10	5

It represents 3 x 100 + 6 x 10 + 5.

In both these systems there is no limit to the number of symbols required. We need a different symbol for millions, another symbol for 10 millions, and so on. The modern system uses precisely 10 symbols: the digits 0 to 9.

The value of each digit is shown by its place in the number. In 365, for example, the digit 5 on the right represents 5, the digit 6 represents 60, as it is one place to the left, and the 3 represents 300. This system came to the West from India via the Arab countries and is known as the Indo–Arabic system.

The ancient Babylonian place value system was even more economical. It used only two symbols: ❙ for 1 and ⟨ for 10. The place value system consisted of grouping numbers in powers of 60 rather than of 10. The following number

❙❙❙ ⟨⟨❙ ⟨⟨⟨⟨❙❙❙

represents $3 \times 60^2 + 21 \times 60 + 43 = 12\ 103$.

Fractions

There are different systems for writing fractions.
This has always been the case, even in ancient times.
For example, the Egyptian system was very limited,
while the Babylonian system is still in use today.

Any advanced civilization has a system of writing fractions. Despite their renowned technological prowess, the Ancient Egyptians had a system of fractions that was comparatively clumsy.

With the exception of $\frac{2}{3}$, the only fractions recognized by the Ancient Egyptians were those with 1 on the top, called aliquot fractions, such as $\frac{1}{2}, \frac{1}{3}, \frac{1}{4}$. Any other fraction had to be written in terms of these aliquot fractions. Furthermore, they were not allowed to repeat a fraction. If they wanted to write $\frac{2}{5}$, for example, they could not write it as $\frac{1}{5} + \frac{1}{5}$. For the second $\frac{1}{5}$, they had to find aliquot fractions with sum $\frac{1}{5}$, such as $\frac{1}{6} + \frac{1}{30}$. So they wrote $\frac{2}{5}$ as $\frac{2}{5} = \frac{1}{5} + \frac{1}{6} + \frac{1}{30}$ (and there are other possibilities too).

Few examples of Ancient Egyptian mathematics survive, although one that does is a leather scroll, dated from about 1650 BC, which contains fractional calculations such as the one earlier.

The Babylonian system was more flexible, following their system of writing whole numbers. Each unit is divided into 60 smaller parts, called minute parts, then each minute is divided into 60 parts, called second minute parts, and this continues with third minute parts and fourth minute parts. This system is still used today for telling the time. We divide an hour into 60 minutes and a minute into 60 seconds. (Seconds are divided into decimal fractions rather than thirds and fourths, however.)

Why was 60 chosen both for whole numbers and for fractions? Most probably because it has so many divisors and, consequently, many fractions terminate.

Consider the fractions $\frac{1}{2}$, $\frac{1}{3}$, $\frac{1}{4}$ up to $\frac{1}{9}$. Using ordinary decimals, four of them, $\frac{1}{2}$, $\frac{1}{4}$, $\frac{1}{5}$, and $\frac{1}{8}$, have a terminating representation. The other four, $\frac{1}{3}$, $\frac{1}{6}$, $\frac{1}{7}$, and $\frac{1}{9}$, have a recurring representation, such as $\frac{1}{3} = 0.3333...$ (the threes go on ad infinitum). Using Babylonian fractions, only $\frac{1}{7}$ does not have a terminating representation.

Nowadays, we have two ways of writing fractions. When 5 is divided by 8, the result can be written either as $\frac{5}{8}$ or as 0.625.

See: *Writing Numbers,* page 8

Quadratic Equations

*A quadratic equation includes the square of the
unknown. Thousands of years ago mathematicians in
Babylonia knew how to solve quadratic equations.*

**The measurement of land has always
been important to any civilization.
To find the area of a square piece of
land you multiply the side by itself,
which is called the square of the side.
The Latin for square is *quadratus*, and
this is where the word quadratic comes
from. There is always a square term.**

Algebraically, a quadratic equation is
of the form:

$$ax^2 + bx + c = 0$$

where *a*, *b* and *c* are numbers.

The solution (in other words the formula
for *x*) is very well known in school
mathematics all over the world.

$$x = \frac{-b \pm \sqrt{b^2 - 4ac}}{2a}$$

This, of course, uses modern algebraic
notation. However, a *method* for solving
quadratic equations has been known
for thousands of years.

A Babylonian clay tablet in the British
Museum in London contains the
solution to the following problem:

*The area of a square added to the
side of the square comes to 0.75.
What is the side of the square?*

The working shown on the tablet
is illustrated on the left of the table
overleaf (*see page 12*). The modern
algebraic equivalent is shown on
the right.

11

Quadratic Equations

Babylonian tablet	Modern notation
I have added the area and the side of my square. 0.75	$x^2 + x = 0.75$
You write down 1, the coefficient	Coefficient of x is 1
You break half of 1. 0.5	Half of 1 is 0.5
You multiply 0.5 and 0.5. 0.25	$(0.5)^2 = 0.25$
You add 0.25 and 0.75. 1	$0.25 + 0.75 = 1$
This is the square of 1	$\sqrt{1} = 1$
Subtract 0.5, which you multiplied	$1 - 0.5 = 0.5$
0.5 is the side of the square	$x = 0.5$

In general, the method gives the following formula to solve the equation $x^2 + bx = c$:

$$x = \sqrt{\left(\frac{b}{2}\right)^2 + c} - \frac{b}{2}$$

This is more or less the same as the modern formula given above, where $a = 1$.

The Greatest Pyramid

*A frustum of a pyramid is a pyramid with its top cut off.
An ancient Egyptian manuscript gives a method for
calculating the volume of this.*

Ancient Egypt is particularly famous
for the construction of the Pyramids.
The engineering skills that went into
their construction have, unfortunately,
been lost to us. Likewise we can now
only guess at the mathematical skills
the Egyptians possessed.

Take a solid like a cone or a pyramid,
which slopes uniformly from its base
to a point at the top. If we cut a slice
off the top the result is a frustum. A
yoghurt pot is an example of a frustum
of a cone.

The Moscow papyrus, dating from
about 1850 BC, contains a set of rules
for finding the volume of a frustum of a
pyramid. It goes:

*Given a truncated pyramid of height
six and square bases of side four on
the base and two at the top.
Square the four, result 16.*

*Multiply four and two, result eight.
Square the two, result four.
Add the 16, the eight and the
four, result 28.
Take a third of six, result two.
Multiply two and 28, result 56.
You will find it right.*

Following these rules, this method
gives a formula for the volume as:

$$^1/_3 \times 6 \, (4^2 + 2 \times 4 + 2^2) = 56.$$

This does give the correct volume.

Generalizing, if the frustum has
height h, a square top of side r and
a square base of side R, the method
gives the following formula for its
volume:

$$^1/_3 h \, (R^2 + Rr + r^2)$$

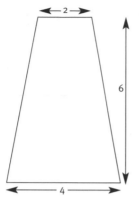 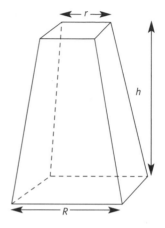

The illustration shows truncated pyramids.

which is correct. No indication is given for how this method was reached. Was it by experiment, or from theory?

This mathematical result was described (by a mathematician, mind you) as the "Greatest Egyptian Pyramid."

π

*The ratio of the circumference
of a circle to its diameter.*

The value of π has been found to higher and higher accuracy. It occurs in many places in mathematics besides the measurement of circles.

Circles come in different sizes of course. As the diameter (the length across) increases, so also does the circumference (the length around). The ratio between these two is the same for all circles and it is given the name π (Greek letter p, pronounced "pie").

All civilizations have needed to find an approximation for π. An early Egyptian value was $4 \times (^8/_9)^2$, which is 3.16, close to 3.14. In the Bible, I Kings 7, verse 23, the more approximate value of three is given.

The first-known reasoned estimation of π is due to Archimedes in the 3rd century BC. By drawing polygons inside and outside a circle, with more and more sides, he was able to close in on the value of π. With polygons of 96 sides, he found that π lies between $^{223}/_{71}$ and $^{22}/_7$. The latter value is still used. In the fifth century, a Chinese mathematician, Zu Chongzhi (429–501), found the more accurate fraction $^{355}/_{113}$.

Further progress was made possible by the development of trigonometry. In the 14th century the Indian mathematician, Madhava, used trigonometry to discover the following series (which continues forever:

$$^\pi/_4 = 1 - {}^1/_3 + {}^1/_5 - {}^1/_7 + \dots$$

This can be used to find π, but it is a very inefficient method. Using a variant of the series Madhava was able to calculate π to 11 decimal places.

Until the 20th century all the calculations were done by hand but with the invention of computers, much greater accuracy is possible. In 1949, the ENIAC calculated π to 2,037 decimal places, taking 70 hours to do so. Modern computers have calculated π to well over a million places.

The number π occurs throughout both pure and applied mathematics. Often these applications have nothing to do with the measurement of circles. For example, the equation of the normal or bell curve, which is central to statistics, is:

$$y = \frac{1}{\sqrt{2\pi}} e^{-\frac{1}{2}x^2}$$

See: *ENIAC,* page 181; *The Normal Distribution*, pages 94–95

The Pythagoreans

The Pythagorean slogan was:
All Things Are Numbers.

The Pythagoreans were a religious, mystical, and scientific sect mainly based in Southern Italy in the 6th century BC.

Their leader, Pythagoras himself, may or may not have existed. Many incredibly important discoveries are credited to the Pythagoreans, of which some will appear in this book.

The Pythagoreans are credited with discovering the following:
• That the Earth is a sphere.
• That the Earth is not the center of the universe.
• That musical harmony depends on the ratio of whole numbers.

No one knows what the Pythagoreans' slogan, *All Things Are Numbers*, means.
Does it just mean that all things can be described in terms of numbers?
Or is it something stronger, that the solid world is an illusion and that the reality behind it consists of numbers?
More important, however, than any single discovery is the Pythagoreans' contribution to mathematics, extending it from a practical subject concerned with areas of land or weights of corn to the study of abstract ideas.

Pythagoras's Theorem

For a right-angled triangle, the square on the hypotenuse is equal to the sum of the squares on the other two sides.

Pythagoras is credited with the proof of this most famous theorem in mathematics.

There are several hundred proofs of the theorem. The visual one below is just one example:
Take a right-angled triangle with sides a, b, and c, where c is the hypotenuse, the longest side. Make four copies of this triangle. Draw a square of side $a + b$. The four triangles are arranged inside the square in two ways. In both cases, look at the region left uncovered by the triangles.

In the upper diagram, the triangles are put in the four corners. The region left uncovered is a square of side c, which has area c^2.

In the lower diagram, the triangles form two rectangles, at the top left and bottom right. The uncovered region consists of two squares, one of side a, the other of side b. The area is $a^2 + b^2$.

The region left uncovered must be the same in both diagrams.
Hence $c^2 = a^2 + b^2$.

The theorem (though probably not its proof) may have been known long before Pythagoras. There are Babylonian clay tablets dating from about 2000 BC, which seem to provide numerical instances of the theorem.

Irrational Numbers

An irrational number cannot be expressed as the ratio of two whole numbers. Many numbers, such as √2, the square root of √ are irrational.

The Pythagoreans thought that everything could be explained in terms of whole numbers and their ratios – fractions, in other words. It was a great shock when it was shown that this is not true.

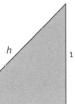

Take a right-angled triangle in which the two shorter sides each have a length of one unit. According to Pythagoras's theorem, the length of the hypotenuse h is given by $h^2 = 1^2 + 1^2$. So h^2 is 2, and hence h itself is √2, the square root of 2. This number is not the ratio of two whole numbers, and hence is an *irrational* number.

The proof is the earliest example of a "proof by contradiction." It assumes that √2 is a rational number, i.e. that $√2 = {}^a/b$, where a and b are whole numbers, and derives a contradiction.

This proof is one of the most important in the history of ideas. It destroyed the notion that everything could be described in terms of whole numbers. The actual person who made the discovery remains unnamed but his fellow Pythagoreans were so appalled by his impudence that they drowned him in the Aegean Sea.

See: *The Pythagoreans,* page 17

Perfect Numbers

*A number is perfect if it equals the
sum of its proper divisors.*

The search continues for perfect numbers, especially an odd perfect number.

Numerology is the magical side of mathematics and some traces – such as perfect numbers – remain in modern mathematics. Perfect numbers were thought to be mystically superior to all others and this can be seen by the following quotation from St Augustine's *City of God* (420 AD):

Six is a perfect number, not because God created the world in six days, rather the other way round. God created the world in six days because six is perfect...

A **perfect** number is equal to the sum of its proper divisors. The first two perfect numbers are 6 and 28.

The divisors of 6 are 1, 2, and 3.

$6 = 1 + 2 + 3$

The divisors of 28 are 1, 2, 4, 7, and 14.

$28 = 1 + 2 + 4 + 7 + 14$

The next perfect numbers are 496 and 8128, the only ones known before the 13th century. The next three were found (along with three incorrect numbers) by Arab mathematician Ibn Fallus.

Finding even perfect numbers is comparatively easy. There is a formula for them, which essentially appears in Euclid's *Elements*. The formula is $2^{n-1}(2^n - 1)$, provided that the term inside the brackets is a prime number.

All the perfect numbers that have so far been discovered are even; an odd perfect number, if it exists, remains to be found. This is the oldest unsolved problem in mathematics.

Certainly there are no odd perfect numbers up to 10^{300} (1 followed by 300 zeros). They may not exist, but if one is ever found, mathematicians will already know a lot about it: that it has at least nine prime factors, for example.

Regular Polygons

A regular polygon has equal angles and equal sides.

Examples of regular polygons are the equilateral triangle (all sides equal, all angles equal to 60°) and the square (all sides equal, all angles equal to 90°). Then comes a pentagon, then a hexagon, and so on.

How do you draw these shapes? Greek mathematicians were very particular about exactness in geometry, and required exact constructions. They would not allow a protractor to measure and draw angles, because one cannot do so exactly. They did not allow a ruler to measure and set out lengths, as one cannot be sure one has the exact length. These constructions had to be made with two instruments only – a straight edge and compasses.

Constructions of an equilateral triangle and a square are part of school mathematics. The construction of a triangle is shown.

The line *AB* is drawn, then arcs of the same length as *AB* are drawn to intersect at *C*. Notice that a straight edge has been used to draw the lines, and compasses to draw the arcs. We do not need to use a protractor to measure an angle of 60°.

With a lot more work, it is possible to construct a regular pentagon. Hexagons and octagons are straightforward. Heptagons (7 sides) and nonagons (9 sides) had to wait!

The diagram shows the construction of an equilateral triangle.

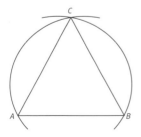

Platonic Solids

There are precisely five Platonic solids.

For a *regular* or *Platonic* solid, all the faces are equal regular polygons.

A regular polygon, such as a square or an equilateral triangle, has equal angles and equal sides. The best-known Platonic solid is the cube, whose six faces are equal squares.

The proof that there are no more than five such solids appears as the very last proposition in Euclid's *Elements*.

They are called Platonic solids from their appearance in Plato's *Timaeus* (dated about 350 BC). This is an obscure and ambiguous book, however, with many possible interpretations. It contains what could be described as an atomic theory in which the four elements of matter – fire, air, water, and earth – consist of these solids. They look as follows:

fire	tetrahedron
air	octahedron
water	icosahedron
earth	cube

Fire, for example, consists of countless atoms, each of which is a tiny tetrahedron. The sharp points of this solid explain why fire is painful.

Earth (or solid matter in general) consists of atoms, each of which is a tiny cube. The fact that cubes can be densely stacked together explains why earth is heavy.

The dodecahedron represents star and planet matter, which was believed to be different from matter on the Earth.

The five solids were known before Plato. They are attributed to the Pythagoreans, who reportedly sacrificed one hundred oxen to celebrate the discovery of the dodecahedron.

See: *The Pythagoreans,* page 17; *Regular Polygons,* page 21; *Euclid's* Elements, page 35.

Tetrahedron: four triangular faces.

Octahedron: eight triangular faces.

Cube: six square faces.

Dodecahedron: 12 pentagonal faces.

Icosahedron: 20 triangular faces.

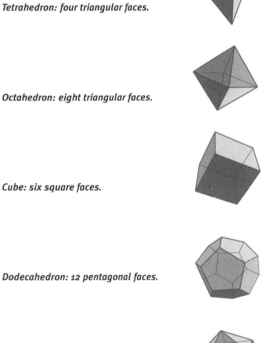

The Golden Ratio

A ratio of lengths that occurs in mathematics, nature, and art.

The sides of a rectangle are in the golden ratio if, when you remove a square, the new rectangle is similar to the original one. The golden ratio is a number that is defined geometrically but which occurs in many other contexts.

The diagram shows the situation. The sides are r and 1. If we remove a square of side 1 by cutting along the dotted line, we now have a rectangle which is 1 by $r - 1$.

This is similar to the original rectangle. Hence the ratios $r/1$ and $\dfrac{1}{r-1}$ are equal.

Putting $r = \dfrac{1}{r-1}$, we obtain:

$$r^2 - r - 1 = 0.$$

The positive solution of this quadratic equation is the value of r.

This ratio is called the golden ratio, or golden section, and it is written as φ (pronounced phi). Like so

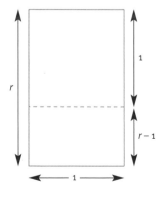

This illustrates how the golden rule is defined.

many other things, its discovery is credited to the Pythagoreans.

The exact value of φ is $\frac{1+\sqrt{5}}{2}$, and an approximate value is 1.618.

Here are some of the occurrences of the ratio:

- In mathematics, it occurs in the pentagon and the pentagram (five-pointed star) and the Penrose tiling. The ratio of successive terms of the Fibonacci sequence tends to φ
- In nature, the shell of the Nautilus snail and the pattern of sunflower petals are said to exhibit the ratio.

- In art, the façade of the Parthenon in Athens is reputed to be a rectangle in the golden ratio though this is controversial. Renaissance artists were very interested in the ratio and it appears in many paintings. Luca Pacioli, the inventor of double-entry bookkeeping, wrote a book on the ratio called *De divina proportione*, illustrated by Leonardo da Vinci. Composers such as Béla Bartok and Claude Debussy deliberately used the ratio in their music.

The ratio of 1 mile to 1 kilometer is 1.609, very close to 1.618, but that is probably just a coincidence.

See: *Fibonacci Numbers*, pages 54–55; *Tessellations*, pages 68–69.

Trisecting the Angle

*The problem of dividing an angle
into three equal parts.*

We can bisect an angle, but can we trisect it?

Greek mathematicians set many problems. This and the next two topics contain the three most important problems, having had a great influence on the progress of Greek mathematics and, indeed, on all mathematics.

All three problems are geometrical. They involve performing a geometrical construction. For Greek mathematicians, constructions had to be exact, using straight edge and compasses only. You were not allowed to use a ruler to measure distances nor a protractor to measure angles.

Suppose we are given an angle. The problem is to trisect it or, in other words, to cut it into three equal angles. Can this be done with straight edge and compasses only? You are not meant to measure the angle with a protractor and then divide by three.

To bisect it (or to cut it into two equal parts) is straightforward. To bisect ∠ ABC:
Put the point of the compasses on B. Draw an arc cutting the lines at D and E. Put the point of the compasses at D, draw an arc. Put the point of the compasses at E, draw an arc. These arcs meet at F. FB bisects the angle.

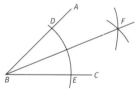

Is there a similar construction to trisect the angle? The Greeks found many clever methods but they all required more instruments than the basic straight edge and compasses.

This matter has been a very popular problem among amateur mathematicians and is possibly one for which the greatest number of false proofs have been proposed.

Doubling the Cube

*Given a cube, construct a cube with
exactly twice its volume.*

The problem of doubling the cube is equivalent to constructing a line with a certain length, using straight edge and compasses only.

In ancient times, the island of Delos in the Aegean Sea was troubled by a plague. When the people consulted an oracle, it told them that the fault lay in the altar to Apollo which was shaped like a cube. The god was offended because it was too small and it should be made twice as large.

The Delians reconstructed the altar, doubling its height, width, and depth, but the plague continued unabated. A new appeal to the oracle revealed the reason for this: the altar was now too large.

Suppose the side of the original cube was 1 unit. As all three dimensions had been doubled, they were each 2 units.

The volume of the altar was now 2 x 2 x 2 = 8 times the original volume. The god Apollo wanted the volume of the altar, not its sides, to be twice as large.

Suppose each side becomes k units. Then the volume of the altar is $k \times k \times k$, which is k^3. If the volume is now doubled, then $k^3 = 2$. Therefore, k itself must be $\sqrt[3]{2}$, the cube root of 2.

The problem now becomes the following: given a length of one unit, construct a length of $\sqrt[3]{2}$ units. Greek mathematicians (and Greek gods, presumably) were very particular about exactness. The Greeks tried all kinds of ingenious methods to construct this length, but all of the methods used something other than straight edge and compasses. It would be many years before the answer to this conundrum was finally found.

See: *Doubling the Cube and Trisecting the Angle Revisited*, page 118

Squaring the circle

This is another of the three Greek problems (the others being Trisecting the Angle and Doubling the Cube): Given a circle, construct a square of equal area.

The problem of squaring the circle reduces to the following: given a line of length 1, construct a line of length π. As always in Greek geometry, the only instruments you are allowed are a straight edge and compasses.

The phrase "squaring the circle" has entered ordinary language to mean a task that is inherently impossible. The original meaning was subtler and less clear cut, however. It means to find a method of constructing a square exactly equal in area to a given circle. It is far from obvious that this is impossible.

Suppose, for simplicity, that the circle has a radius of 1 unit. Then its area is $\pi \times 1^2$, which is π. If the equivalent square has side x units, then its area is x^2.

The problem now becomes to find a length x such that $x^2 = \pi$. The side x of the square must be equal to $\sqrt{\pi}$, the square root of π.

The square root is not a problem. If you can construct a length k, then you can construct a length $\sqrt{k}$. The problem is π. Ingenious methods were invented to draw a line of length π, but they all used moving parts, or required curves that could not be drawn exactly, such as spirals.

The three Greek problems have all been resolved, but only over 2,000 years since they were originally posed.

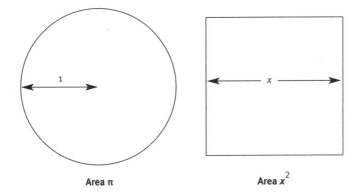

Area π   **Area x^2**

The above illustration shows a circle and square of equal area.

See: *Squaring the Circle Revisited*, page 133

Zeno's Paradoxes

Zeno (c. 490–c. 430 BC)

These paradoxes concern the infinite divisibility of space and time, and suggest that motion is impossible.

Matter cannot be divided indefinitely. There are particles that cannot be cut up any further. Is the same true of space and time, or can they be divided indefinitely? Zeno's paradoxes concern this question.

Zeno's paradoxes include the following two puzzles:

1. For the Dichotomy, suppose that you want to cross a field. Before you reach the other side you must get half way across. Before you reach the halfway point you must get a quarter of the way across and so on. To cross the field you must travel an infinite number of smaller distances, and so it is impossible to get across at all.

2. Achilles and the Tortoise have a race. Achilles can run ten times as fast so the Tortoise is given a lead of 100 paces. By the time Achilles has run 100 paces the Tortoise is 10 paces ahead. By the time that Achilles has made up these 10 paces the Tortoise is one pace ahead, and so on. Achilles can never catch up with the Tortoise.

Both these paradoxes cut space up into infinitely many pieces. In mathematics it is possible for infinitely many numbers to have a finite sum. This was shown by Archimedes in about 210 BC and formed an important part of mathematics in the 17th century. The two paradoxes can be written thus:

The Dichotomy:

$$\tfrac{1}{2} + \tfrac{1}{4} + \tfrac{1}{8} + \tfrac{1}{16} + \ldots = 1$$

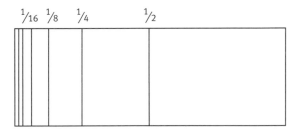

Is it possible to cross the field?

Achilles and the Tortoise:

$$100 + 10 + 1 + \frac{1}{10} + \frac{1}{100} + ... = 111\frac{1}{9}$$

Therefore, Achilles will catch up with the Tortoise after running $111\frac{1}{9}$ paces – a finite distance.

Although this explanation satisfies mathematicians, it does not answer the question of how it is possible to accomplish infinitely many tasks in a finite period of time. Zeno thought it impossible and hence that any sort of motion is an illusion.

Plato and Platonism

Plato (428–348 BC)

A philosophy of mathematics that claims that mathematics exists outside the human mind and that it is essential for the education of enlightened people.

Plato was one of the most important philosophers of all time. His name has even entered our language: a "platonic" friendship is one without any sexual content.

He was also a mathematician and held the discipline of mathematics in very high regard. Above the gate of his academy was written: *No one ignorant of geometry can enter here.*

Plato's own achievements in this field were minor – one (disputed) story is that he invented a device with moving rods for the problem of "doubling the cube." On the other hand, his influence on the philosophy of mathematics was enormous.

In mathematics, Platonism is the theory that mathematical objects – numbers, triangles, and so on – have an existence independent of the human minds that think about them. It is a theory that is very difficult to justify without extra philosophical assumptions. Where do these abstract ideas exist? Is there another universe, completely different from our material universe, which contains abstract objects? Plato seems to have thought so. The essence of his philosophy is that the material world is but an inferior copy of the world of abstract forms.

According to Plato, the work of the mathematician involves discovery rather than invention. Mathematicians investigate the universals which exist independently of mankind, rather than create ideas from their own minds.

The other justification is theological. Abstract ideas exist in the mind of God. Plato said: "*In mathematics, men think the same thoughts as gods.*"

Conic Sections

Menaechmus (380–320 BC)

*These are a set of curves that are
exposed when a cone is cut.*

Conics occur in many places in nature
as well as in mathematics.

Imagine a double cone (similar to the
one in the diagram below). This cone is
cut by a plane. The exposed surface is a
conic section.

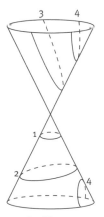

Double cone.

The type of conic depends on the
angle of the plane. The numbers in the
diagram indicate the following:

1. The plane is horizontal: a circle.
2. The plane is slightly tilted: an
   ellipse.
3. The plane is parallel to the side of
   the cone: a parabola.
4. The plane is steeply tilted: a
   hyperbola. Notice there are two
   parts to this curve, both labeled
   number 4.

There are many ways of defining
conics without involving a three-
dimensional cone. Take a fixed point
F and a fixed line *d* as in the diagram
shown on the following page. Suppose
a variable point *X* moves so that the
ratio *XF*:*Xd* is constant. Then *X* moves
in a conic.

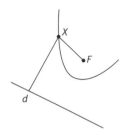

Focus and Directrix.

The ratio is called the eccentricity, the value of which tells us the sort of conic that is defined. The fixed point *F* is called the focus. The Earth moves round the Sun in an ellipse with the Sun at the focus. The eccentricity is about $\frac{1}{90}$.

There are many other real-life examples of conics. A wheel seen at an angle is an ellipse. Cut a cucumber, and the exposed surface forms an ellipse. Throw an object in the air and you will see that its path is a parabola. The reflecting surface of a car headlight is formed from a parabola.

The Long Range Navigation (LORAN) system used to guide ships involves intersecting hyperbolas.

Euclid's *Elements*

Euclid (c. 325–265 BC)

These are a compilation of geometric theorems.

Euclid's *Elements* lists and proves the geometrical results of his day. It has survived as a model of logical reasoning and as a basis for learning mathematics for 2,000 years.

Euclid's *Elements of Geometry* consists of 13 books that cover the mathematics known during the time of writing. It is largely a compilation of other mathematicians' results rather than original work by Euclid himself, but its structure, a chain of theorems proceeding logically from clearly stated definitions and axioms, is due to Euclid.

The collection contains all traditional school geometry. The fifth proposition is that the base angles of an isosceles triangle are equal, the 32nd is that the sum of the angles of a triangle is 180° and the 47th is Pythagoras's Theorem.

It is very much *pure* mathematics. When a pupil at Euclid's academy asked what gain he could make from it, Euclid contemptuously told his slave to throw the student a coin.

The theorems are all phrased in terms of geometry, as was nearly all Greek mathematics, but a large part of the work lays the foundation for number theory.

The book was used continuously for teaching mathematics for 2,000 years. British intellectual Bertrand Russell was introduced to the *Elements* at the age of 11 and recorded the impression it made on him:

This was one of the great events of my life, as dazzling as first love. I had not imagined there was anything so delicious in the world.

The Fifth Postulate

Euclid (c. 325–265 BC)

The postulate, proposed by Euclid, which gives the properties of parallel lines.

———————

Euclid's *Elements* consists of a chain of theorems each proceeding from the ones before. The starting point of this chain is a set of axioms and postulates, one of which is far from obvious.

Any logical chain of reasoning must start from somewhere, and in the case of Euclid it is a set of common notions and postulates. The common notions apply to all reasoning and are uncontroversial. They contain obvious statements such as the first one which states: *Things which are equal to the same thing are equal to each other.*

The postulates are specifically about geometry. The first four are deemed unexceptionable, for example the fourth states: *...all right angles are equal to each other.*

The fifth postulate is much longer and more complicated than the others:

That, if a straight line falling on two straight lines make the interior angles on the same side less than two right angles, the two straight lines, if produced indefinitely, meet on that side on which are the angles less than the two right angles.

In the diagram on the opposite page, *XY* and *PQ* are the two straight lines. Another straight line crosses them, and the interior angles on the same side are *a* and *b*.

Suppose that the sum of these angles is less than two right angles, that is, $a + b < 90° + 90° = 180°$. Then, if *XY* and *PQ* are extended in both directions, they will meet on the left-side of the diagram.

This is also called the parallel postulate. If *XY* and *PQ* are parallel,

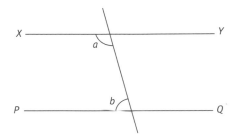

then they will never meet on either side. Hence $a + b$ must be exactly 180°.

Euclid manages to avoid this postulate until Proposition 29, which proves well-known results about alternate and corresponding angles. He has to use it to prove, for example, that the sum of the angles of a triangle is 180°. The fifth postulate is also necessary to prove Pythagoras's Theorem and many other standard results of geometry.

There were many attempts to prove the fifth postulate from the other four but these attempts were always shown to contain some other assumption.

It was a long time before people began to suspect that it was impossible to prove the fifth postulate and that there are many possible geometries, some assuming the fifth postulate and some denying it.

See: *Non-Euclidean Geometry*, page 113–114

Sum of the Angles in a Triangle

*The sum of the angles in a triangle
is two right angles, or 180°.*

For any triangle, the sum of its angles is always 180°. This result relies on a disputed axiom of geometry and is familiar from school geometry.

Take a triangle *ABC* and draw a line *DAE* parallel to *BC*.

Then $\angle ABC = \angle DAB$ and
$\angle ACB = \angle EAC$.
So $\angle BAC + \angle ABC + \angle ACB$
$= \angle BAC + \angle DAB + \angle EAC$.

The left-hand side of this equation, $\angle BAC + \angle ABC + \angle ACB$, is the sum of the angles of the triangle.

The right-hand side, $\angle BAC + \angle DAB + \angle EAC$, is the sum of the angles along a straight line, which is 180°. So the sum of the angles of the triangle is 180°.

In the proof above we stated that:

$\angle ABC = \angle DAB$ and $\angle ACB = \angle EAC$

These pairs of angles are known as "alternate angles". The equality of alternate angles is a consequence of the fifth postulate of Euclid concerning parallel lines. A different postulate could give a different result: it might be that the sum of the angles is less than 180°, or greater than 180°.

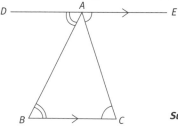

Summing the angles.

The Fundamental Theorem of Arithmetic

Euclid (c. 325–265 BC)

Every whole number can be written as a product of prime numbers.

It is possible to factorize any whole number until one is left with prime numbers which, by definition, cannot be factorized any further.

A prime number is a whole number whose only factors are 1 and itself. For example 11 is a prime number, as it cannot be written as the product of smaller numbers. The number 15 is not a prime number, as it can be written as 3 x 5. The first few prime numbers are 2, 3, 5, 7, and 11. Prime numbers are of supreme importance in the theory of numbers.

Many branches of mathematics have a "fundamental theorem." Often this is an arbitrary choice but when there is a result from which all the other results flow, it is natural to choose it as the fundamental one. In the case of arithmetic, the fundamental theorem says that you can factorize whole numbers into prime numbers. Furthermore, for each number there is only one possible factorization.

For example:

12 = 2 x 2 x 3
35 = 5 x 7
1001 = 7 x 11 x 13

To draw an analogy with chemistry, prime numbers are like atomic particles, that is, they cannot be split up and every other number can be expressed in terms of them.

The Infinity of Prime Numbers

Euclid (c. 325–265 BC)

The number of primes is infinite.

However many prime numbers are written down, there will always be another one.

A prime is a number which has exactly two divisors – 1 and itself. The first few prime numbers are 2, 3, 5, 7, 11, and 13.

This sequence goes on for ever. The proof is in Euclid's *Elements*. His proof shows that there are more than three primes, but it can be extended to any number.

The proof goes by contradiction. Suppose that the only primes are *a*, *b*, and *c*. Then consider *abc* + 1. This is one greater than a multiple of *a*, and so it cannot be divisible by *a*. Similarly it cannot be divisible by *b* or *c*. By the fundamental theorem of arithmetic, any number can be written as a product of primes. This new number must be divisible by a prime which is different from *a*, *b*, and *c*, contradicting the assumption that these are the only primes.

This proof shows that there are infinitely many primes but it does not provide a formula for listing them. That formula was still to be discovered, over 2,000 years in the future.

See: *Euclid's* Elements, page 35; *The Fundamental Theorem of Arithmetic*, page 39

Measurement of a Sphere

Archimedes (287–212 BC)

*This concerns formulae for the volume
and surface area of a sphere.*

Archimedes is often described as the
archetype of the absent-minded
professor. Yet, he achieved more in
mathematics, science, and technology
than anyone else in the ancient world.

The diagram shows a sphere inside a
cylinder. From this diagram Archimedes
found two things:

1. The surface area of the sphere is the
   same as that of the cylinder
   (without the ends).
2. The volume of the sphere is two-
   thirds that of the cylinder.

The formulae $4\pi r^2$ for the surface
area, and $^4\!/_3\,\pi r^3$ for the volume
followed. Archimedes was so proud of
these results that the diagram was
engraved on his tombstone.

Archimedes is famous for mathematics,
scientific discoveries concerning centers
of gravity and floating bodies, and such
inventions as the Archimedes screw. He is
also famous for leaping out of his bath
and running around shouting "*Eureka!*".

While the Roman army besieged
Syracuse, where Archimedes was living,
he invented terrifying war machines to
drive them back. Despite this, they finally
took the city. While he was busy tracing a
geometrical figure on the sand, a soldier
summoned him to attend the Roman
governor. "*Don't disturb my circles!*" he
protested. These were his last words.

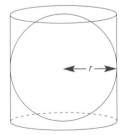

Sphere inside cylinder.

Quadrature of the Parabola

Archimedes (287–212 BC)

This is about finding the area between a chord and a curve.

To evaluate the area between a chord and a curve, Archimedes found a way to add infinitely many numbers.

A parabola is an example of a conic which occurs in many places in science as well as mathematics. The reflecting surface of a space-telescope, such as that at Jodrell Bank in Manchester, England, is formed from a parabola.

The diagram (*page 43*) shows part of a parabola curve and a straight line crossing it. Archimedes found the area of the shaded region, between the line and the curve.

This area is $^4/_3$ of the area of the triangle with the same base and height. Although this was not a terribly interesting result, it was the method of proof that was ground breaking. Put one triangle in the shaded region, which leaves two gaps. Put two triangles in the

gaps, leaving four gaps. Repeat indefinitely; the area of all these triangles approaches the area required. The first two stages are shown.

The method of proof is known as the method of exhaustion. The infinite succession of triangles "exhausts" the area between the curve and the line.

Greek mathematicians distrusted any infinite process. Zeno's paradoxes are about the adding of infinitely many terms and obtaining something finite. Archimedes showed that it was possible.

To find the area, Archimedes had to sum infinitely many smaller areas. He showed that this sum of infinitely many fractions:

$$^1/_1 + {}^1/_4 + {}^1/_{16} + {}^1/_{64} + {}^1/_{256} + \dots$$

has a finite value, $^4/_3$.

This is the first recorded example of the summation of an infinite series, a

Area between line and curve

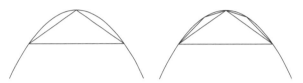

"Exhausting" the area

key part of mathematics, and moreover it was found rigorously. The formal rigor of Greek mathematics was not achieved again until the 19th century.

It was rumored that Archimedes had a secret method to find these results. In 1906 a palimpsest (a document hidden under another, when the parchment was recycled) was found in a monastery in Constantinople (modern-day Istanbul).

It contained *The Method*, a lost work by Archimedes, which showed how he had first obtained his results via informal reasoning. Essentially, it was the same as the integral calculus of Newton and Leibnitz. In 1998, Christie's auction house in New York sold the manuscript for 2 million dollars to an unidentified collector in the United States.

See: *Zeno's Paradoxes*, pages 30–31; *Conic Sections*, pages 33–34; *Integration*, pages 86–87

The Sand Reckoner

Archimedes (287–212 BC)

Archimedes asked the question:
How many grains of sand would fill the universe?

To answer this, Archimedes had to invent a way of writing numbers much larger than any used before.

At the time, the universe was thought to be finite in radius, being bounded by the sphere of the stars. In the "Sand Reckoner," Archimedes made estimates for both the size of the universe and of a grain of sand, and had to find how many of the latter would fit into the former.

The problem was that at the time no notation existed to express such a huge number. The largest number word the Greeks had was "the myriad," which means 10,000. They also used "the myriad myriad," in other words, 10,000 x 10,000, or a hundred million, or 10^8 in modern notation.

In ordinary notation, we go up in steps of 10, then 100, then 1,000, and so on. Single-digit numbers are less than 10,

two-digit numbers are less than 100, and three-digit numbers are less than 1,000. Archimedes extended this principle, replacing 10 by 10^8. He called all numbers up to 10^8 numbers of the first order. Using 10^8 as a starting point, he took successive multiples of this new unit. He called all numbers between 10^8 and $10^8 \times 10^8 = 10^{16}$ numbers of the second order. This is continued to numbers of the third order and so on, ending with numbers of the myriad myriadth order, which starts at $(10^8)^{10^8}$.

Now Archimedes had his number system and he concluded that the number of grains required is 10,000,000 units of the eighth order, which is 10^{63}.

This anticipates much of modern ways of writing large numbers. It also includes a tantalizing reference to the claim of another Greek scientist, Aristarchus, that the Earth travels round the Sun.

Trigonometry

Hipparchus (190–120 BC)

Trigonometry is concerned with calculating sides of triangles from angles and early development was mainly used in astronomy.

Trigonometry is taught throughout high schools. It is based on three functions: sine, cosine, and tangent.

The original trigonometric function was the chord function. Start with an isosceles triangle rather than a right-angled one. Let the equal sides each have length one unit. The chord function,

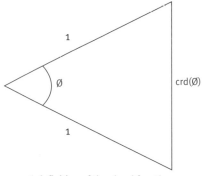

A definition of the chord function.

crd(Ø), gives the third side of the triangle.

It is easy to convert between the chord and sine functions, with these formulae, which involve only doubling and halving:

$$crd(Ø) = 2\sin(\tfrac{1}{2}Ø) \quad \sin(Ø) = \tfrac{1}{2}crd(2Ø)$$

No trigonometrical work of Hipparchus survives but he is known to have compiled the first trigonometric table – a table of values of the crd function.

The oldest surviving table is in the *Almagest* of Ptolemy, a work of astronomy. The table is a feat of numerical complexity: starting with results like crd(60°) = 1 and crd(90°) = √2, and using formulae for crd(A + B) and crd(A − B), the chords of angles are found for every $\tfrac{1}{2}$°, to an accuracy of up to 6 decimal places.

The familiar functions of sine, cosine, and tangent, introduced by Indian and Arab mathematicians, are now used but the methods remain the same.

Negative Numbers

The extension to numbers less than zero.

Negative numbers make sense in some contexts, but not in others. **Where they are relevant they save a lot of time but mathematicians did not accept them as proper numbers until comparatively recently.**

It makes no sense to say *"There are minus five people in the room."* However, in many other contexts it is useful to have numbers which are less than zero. One familiar example is temperature: at 0°C water freezes and we require numbers to describe temperatures lower than that figure. In commerce, too, it is useful to have negative numbers to describe a debt.

Chinese mathematicians were the first to accept negative numbers. They did their arithmetic on a chequerboard using short rods for the numbers. Red rods were used for positive numbers and black for negative, whereas the modern way of describing whether a bank-balance is positive or negative is the opposite way round.

Greek mathematicians did not recognize negative solutions of equations. In the 3rd century, Diophantus rejected $x + 10 = 5$, saying it was not a proper equation. Indian mathematicians came closer to accepting negative numbers, finding negative roots of quadratic equations. However, Bhaskara II (1114–1185) the leading mathematician in the 12th century, rejected these solutions, stating that people did not approve of them.

Nowadays, negative numbers are an essential part of mathematics. An example of their usefulness is in solving quadratic equations.

The equation $ax^2 + bx + c = 0$ is solved by the formula:

$$x = \frac{-b \pm \sqrt{b^2 - 4ac}}{2a}$$

Here a, b, and c can be positive or negative and the same formula fits all cases.

If we do not allow negative numbers, a, b, and c must be positive. There are several separate cases to consider:

$$ax^2 + bx = c$$
$$ax^2 + c = bx$$
$$ax^2 = bx + c$$

Each of these separate cases will have a different formula. That makes three formulae to remember instead of just one!

The product of two negative numbers is positive. This fact is a part of school mathematics that is famously difficult to justify. Teachers have to rely on the following:

Minus times minus equals a plus.
The reason for this we shall not discuss.

See: *Quadratic Equations,* pages 11–12

The Earth-Centered Universe

Claudius Ptolemy (83–161 AD)

This is a system of the universe.

In Ptolemy's system the Earth is at the center and all other bodies revolve around it.

The earliest model for the motion of heavenly bodies about the Earth involved their moving in circles with the Earth at the center. More accurate observations showed that this was incorrect and modifications had to be made. These suggested that:

1. The Earth is not the center of the circle.
2. The bodies move in smaller circles (called epicycles) which were themselves moving in circles around the Earth.
3. The speed of the body was not constant as it moved around the circle.

In Ptolemy's *Mathematike Syntaxis*, the mathematical compilation, better known by its Arabic name *Almagest*, all these modifications are used.

The diagram on the opposite page illustrates this model. A planet moves in a small circle. The center of this small circle moves in a big circle around a point *C* – this is not the Earth – and the angular speed is constant, not about the Earth and not about the center *C* of the big circle, but about a point called the equant *E* – on the other side of the Earth from *C*.

This system was very complicated. It was also reasonably accurate, however, and lasted as a guide for navigation, astronomy, calendar setting, and so on, for well over 1,000 years. Its accuracy was not challenged until the insistence on circular motion was abandoned and other curves were considered.

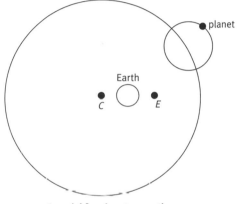

A model for planetary motion.

See: *The Sun-Centered Universe,* page 60;
The Sun-Centered Universe Again, pages 64–65

Zero

Brahmagupta (598–665 AD)

The recognition that zero is a proper number.

Nowadays zero is an ordinary part of how we write numbers and calculate with them. It took a long time for it to be accepted as a proper number.

For most of recorded history the number system started with one. There was no number zero (0) when the system for dating years was devised in 525 AD. The system goes straight from 1 BC to 1 AD – there is no 0 AD.

There is no need for zero when using Roman numerals. The number 105 is *CV* (100 + 5) and we do not have to show there are no tens. By contrast, in a place value system, some indication is required.

Before modern times, arithmetic was done using a counting board, with one column for units, one column for 10s, one column for 100s, and so on. A symbol for zero isn't required – just a space in the relevant column for the missing digit.

When writing numbers, it is useful to have a special symbol to show that a number is missing – such a symbol was used in Central America by indigenous peoples centuries ago. The Babylonians used two small wedges in their sexagesimal system.

Using a symbol for a missing digit in a place value system is not the same as recognizing zero as a number in its own right. That was invented by Indian mathematicians. In 628 AD Brahmagupta wrote out a set of rules for zero including:

• The sum of a zero and a positive number is positive.
• Zero divided by zero is zero.

Now we agree with the first but not the second. Division by zero is forbidden. If a car travels zero miles in zero hours, what is its speed? It is undefined.

Kitab wa al jabr wa al muqabalah (The Book of Shifting and Balancing)

Al-Khwarizmi (c. 780–c. 850 AD)

A comprehensive guide to solving quadratic equations.

Al-Khwarizmi's book describes six types of quadratic equation and solves them in a methodical manner.

Kitab wa al jabr wa al muqabalah means: "The Book of Shifting and Balancing." With an equation you *shift* terms from one side to another, and then you *balance* the terms collected on one side.

Say *Al jabr* quickly. This book is where the word algebra comes from. It is a long way from the algebra studied in schools today: for example, there is no use of letters such as *x* to stand for an unknown.

Indeed, equations are written out in terms of words rather than symbols.

Al-Khwarizmi considers six separate types of quadratic equation, though nowadays the single quadratic formula covers them all. In his time there were no negative numbers, and so $x^2 - 5x = 6$ had to be treated differently from $x^2 + 5x = 6$.

The book contains no startling new discoveries. Its virtue consists in its systematic collection of different results and its methodical treatment of them, in an algebraic rather than a geometric way.

See: *Quadratic Equations*, pages 11–12

Cubic Equations – Geometric Solution

Menaechmus (380–320 BC)
Omar Khayyam (1048–1131 AD)

A geometric method of solving cubic equations.

A cubic equation involves x^3. They were solved by finding the intersection of certain shapes.

A quadratic equation involves the square of the unknown, for example:

$$2x^2 + 3x - 1 = 0.$$

The highest power of x is x^2.

The next step up from the quadratic is the cubic equation, which has an x^3 term as well as the others.

An example of a cubic equation is:

$$3x^3 - 4x^2 + 2x - 7 = 0.$$

In general, a cubic equation in x is of the form:
$$ax^3 + bx^2 + cx + d = 0$$

where a, b, c, and d are known constants, define x. It is considerably harder to solve than the quadratic.

A simple cubic is $x^3 - 2 = 0$. It arose from the problem of "doubling the cube." Menaechmus is recorded as having solved the problem of cubic equations by drawing two parabolas and finding where they intersect.

The diagram shows the parabolic curves $y = x^2$ and $x = \frac{1}{2} y^2$. The value of x where they intersect is $\sqrt[3]{2}$, the solution of $x^3 - 2 = 0$.

Menaechmus and other Greek mathematicians solved cubic equations by intersecting curves, but it seems they were interested in solving specific geometrical problems rather than finding the root of a general algebraic equation.

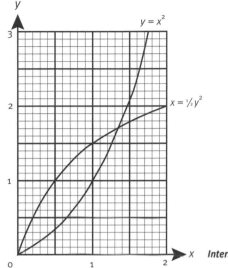

Intersecting parabolas.

The Islamic astronomer and mathematician Omar Khayyam (1048–1131) approached the problem in an algebraic way. Omar Khayyam is best known as a poet. *The Rubaiyat*, a collection of four-line verses, is famous all over the world.

A book of verses underneath the bough,
A jug of wine, a loaf of bread – and thou

Beside me singing in the wilderness.
Oh, wilderness were paradise enow!
– Verse XII, The Rubaiyat of Omar Khayyim

Although these poems are largely concerned with wine and love and have little to do with mathematics, Khayyam is important mathematically as he systematically classified cubic equations and showed how conic curves could be used to solve them.

See: *Doubling the Cube*, page 27; *Conic Sections*, pages 33–34; *Quadratic Equations*, pages 11–12

Fibonacci Numbers

Leonardo of Pisa (Leonardo Fibonacci; 1170–1250)

The Fibonacci sequence of numbers is defined so that each term is the sum of its two predecessors.

This famous sequence of numbers, which has many applications, was devised by the medieval Italian mathematician Leonardo Fibonacci.

In 1202 Fibonacci described his sequence of numbers in terms of breeding rabbits. Start with one adult male/female pair, and assume that:

- Each pair becomes mature after one month.
- Each mature pair produces one new pair each month.
- Rabbits never die.

At the start there is just one pair:

One pair
In month one the first pair breeds.
Two pairs
In month two, the first pair breed again.
Their offspring are still immature.
Three pairs
In month three, the first pair breed, their first offspring breed.
Five pairs
In month four, the first pair and the first two sets of offspring breed.
Eight pairs
... And so it continues.

In the next month there will be 5 + 8 = 13 pairs;
in the month after that 8 + 13 = 21 pairs and so on.

Each term in the sequence is found by adding the previous two terms. The sequence is:

1   2   3   5   8   13   21   34 and so on.

These numbers occur in many contexts. The ratio of successive terms:

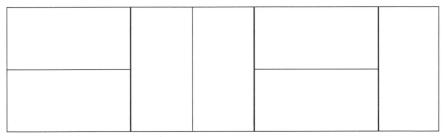

How many ways can the stones be laid?

$$\frac{3}{2} \quad \frac{5}{3} \quad \frac{8}{5} \quad \frac{13}{8} \quad \frac{21}{13}$$

tends to the golden ratio.

Consider the problem of paving a path, 2ft feet wide, with paving-stones that are 1ft by 2ft. You can either lay stones across the path (crosswise) or along the path (lengthwise). If the path is n feet long, the number of possible arrangements is the nth Fibonacci number. One arrangement for $n = 7$ is shown.

The reason is as follows: At stage n, you can either add a crosswise stone to an arrangement $n - 1$ft long, or add two lengthwise stones to an arrangement $n - 2$ft long. Therefore the number of arrangements at stage n is found by adding the number of arrangements at stages $n - 1$ and $n - 2$.

The Fibonacci numbers frequently occur in nature as well. The number of petals in many plants follows the Fibonacci sequence.

See: *The Golden Ratio, 24–25*

Perspective

The geometry needed to draw objects in depth.

A painting represents three dimensions on a two-dimensional surface. The mathematics of perspective gives the painting depth.

In a picture, solid objects are shown on a flat surface. The visitors to a gallery will view the picture from different angles and different distances, and each will want to find it realistic. It is far from clear how this should be done. There is no perfect solution.

Using perspective, during the Renaissance (literally meaning "rebirth," the period following the Middle Ages, in which Europe experienced a new interest in the material world, nature, art, and culture), Italian painters, such as Filippo Brunelleschi (1377–1446), tried to make their pictures as realistic as possible.

Imagine straight railway tracks on a flat plain. They seem to end at a point on the horizon. This point is called a "vanishing point" (a phrase coined by the English mathematician, Henry Brooke Taylor, 1865–1731), although it was an idea raised by the Italian mathematician Leone Battista Alberti (1404–1472) in *De Pictura* (1435).

In one-point perspective, there is one set of parallel lines which meets a single vanishing point. This perspective is useful for drawing a corridor, seen from one end, but for not much else.

In two-point perspective, there are two sets of parallel lines and hence two vanishing points. For example, on a floor with square tiles there are two sets of parallel lines and hence two vanishing points.

Renaissance artists often painted scenes with tiled floors, regardless of historical accuracy, in order to give the impression that distant objects were far away. A diagram on the page opposite shows three-point perspective. Notice the three vanishing points.

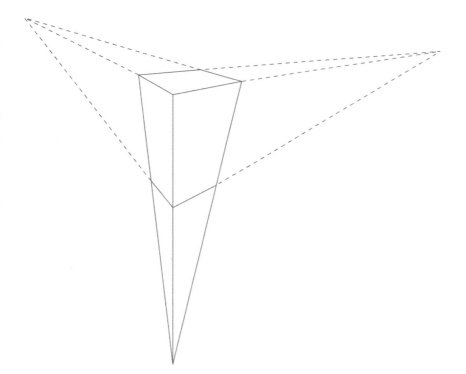

Three-point perspective.

In computer games, a three-dimensional scene or image is depicted on the two-dimensional screen of the monitor. Complicated mathematics of perspective are often used to ensure that the picture, for example, of the dragon's dungeon or the witch's kitchen, does have depth, so that the game player can travel deeper and deeper into peril.

Cubic Equations – Algebraic Solution

Scipione del Ferro (1465–1526)
Niccolo Tartaglia (1500–1557)

An algebraic method of solving cubic equations.

Cubic equations had been solved by geometric methods but later on a purely algebraic solution was found.

Quadratic equations can be solved by the well-known formula. This formula gives the solution in terms of a square root.

The cubic equation involves x^3 as well as x^2 and so on. It was solved geometrically by Greek mathematicians and Omar Khayyam. An algebraic solution would involve cube roots as well as square roots, and Khayyam doubted that such an algebraic solution existed.

A formula giving the solution does exist but is much more complicated than for a quadratic. The general cubic equation can be written in the following form:

$$x^3 + bx^2 + cx = d.$$

An example is $x^3 + 2x^2 + 3x = 4$.

Scipione del Ferro had a method for solving equations in which there was no x^2 term, in other words, of the form $x^3 + cx = d$. He never published it, confiding only in his student Antonio Fiore.

Tartaglia learned how to solve these equations and also those in which there is no x term, such as those of the form $x^3 + bx^2 = d$.

In those days there were problem-solving competitions and so methods of solution were a valuable trade secret. Fiore challenged Tartaglia to such a competition: Fiore set problems of the form $x^3 + cx = d$, which Tartaglia could solve, while Tartaglia's problems were of the form $x^3 + bx^2 = d$, which Fiore couldn't, and so Tartaglia won easily.

Quartic Equations

Lodovico Ferrari (1522–1565)

The solution of quartic equations, which contain a term of the fourth power.

Shortly after cubic equations were solved a method for quartic equations was found.

Ferrari came from a family that had fallen on hard times: at the age of 14 his father was killed and he was forced to become a servant. By good fortune, however, his master was Gerolamo Cardano (1501–1576) who spotted the boy's intelligence and taught him mathematics.

In a quadratic equation the highest power of x is x^2.

In a cubic equation the highest power is x^3.

One step up from the cubic is the quartic, in which there is an x^4 term.

If we collect everything on the left of the = sign, a general quartic equation is of the form:

$$ax^4 + bx^3 + cx^2 + dx + e = 0.$$

While still a teenager, Ferrari found the trick to extend the solution to the cubic to a solution to a quartic. Quite often, mathematicians achieve a breakthrough before they are 20 years old.

Cardano published the result in 1545, alongside Tartaglia's solution of the cubic. The latter was furious and in 1548 another public mathematical debate took place, between Tartaglia, the veteran of 48, and Ferrari, the young man of 26. At the halfway stage Ferrari was winning and so Tartaglia slunk away before the end.

After this victory Ferrari received many offers of employment and was extremely wealthy by the age of 40. He did not live very long to enjoy his success, however, as he was poisoned by his sister a few years later.

The Sun-Centered Universe

Nicolaus Copernicus (1473–1543)

A system of the universe in which the Sun is at the center and all other bodies revolve around it.

It seems obvious, from what we see and feel, that the Earth is stationary and that the heavenly bodies move around it. It requires great imagination to believe that the Earth is in motion, both spinning about its axis and whirling around the Sun.

There were many precedents for a stationary Sun, but the first fully worked-out system was due to Copernicus. It explained several things that are unclear in the Earth centered system: for example, why the planets Mercury and Venus never stray far from the Sun and why some planets seem to go backwards at certain times of the year.

There were disadvantages, though, and it was many years before the system was fully accepted. For a start, people found it hard to believe that the solid Earth was whizzing around in space. Another objection is the phenomenon known as "stellar parallax:" if the Sun and stars are stationary, then the position of the Earth relative to the stars should vary between winter and summer. The measurement of a star from Earth should be different at different times of the year. It isn't, or at least 16th-century measuring equipment was not able to detect any difference. Copernicus's solution was to say that the stars were much further away than had been previously thought. Not until 1838 were scientific instruments sensitive enough to measure stellar parallax.

Copernicus's theory was made public in his book *De Revolutionibus Orbium Coelestium* (*On the Revolutions of the Heavenly Spheres*), which is often credited as the beginning of the Scientific Revolution. Copernicus was terminally ill when it was printed and legend says he got his first copy on the day of his death.

Mathematical Induction

Francesco Maurolico (1494–1575)

A method of proving theorems about whole numbers.

If you can prove something for *n* = 1, and if you can show that you can extend the proof from *n* to *n* + 1, then you have proved it for all *n*.

Mathematical induction is often explained using the basis of a row of dominoes standing on end.

Notice that if one domino falls, the next domino will also fall. Suppose the leftmost domino is toppled. That causes the second domino to fall, which causes the third to fall and so on. All the dominoes will fall in sequence.

Mathematical induction is not to be confused with scientific induction which consists of finding a general law from several specific cases. Mathematical induction is a way of proving results about the positive integers 1, 2, 3, 4, and so on. It is a deductive method.

The method was used implicitly by Greek, Indian and Arab mathematicians, but it was first stated formally by Maurolico. Using a statement S about integers, the method contains three steps:

1. Show that S is true for *n* = 1.
2. Show that if S is true for *n* = *k*, then it is also true for *n* = *k* + 1.
3. State that S is true for all positive integers.

Looking at the domino diagram left, Step 1 corresponds to toppling the first domino. Step 2 corresponds to placing the dominoes so close that if one falls the next falls. Step 3 corresponds to saying that all the dominoes will fall.

By Step 1, we know that S is true for *n* = 1. Use Step 2, putting *k* = 1. Then we know that S is true for $n = 1 + 1$, i.e. $n = 2$. Use Step 2 again, putting $k = 2$. Then S is true for $n = 2 + 1$, i.e. $n = 3$. The argument using Step 2 can be continued up to any positive integer. We can conclude Step 3, claiming that S is true for all positive integers.

The method of induction must involve precise mathematical statements. The following, for example, is not a correct use of the method:

1. *One chocolate won't make me fat.*
2. *Just one more chocolate won't make me fat.*
3. *Therefore I can eat as many chocolates as I like and I won't get fat.*

Falling Bodies

Galileo Galilei (1564–1642)

Rules for how a body falls under gravity.

If a body is dropped from rest the distance it falls varies with the square of the time.

Suppose a body is dropped from a height. It does not fall at a constant speed; rather it picks up speed the farther it falls. Galileo's formula predicts that the distance it falls increases with the square of the time, so in the first second it falls 5m, in the first 2 seconds it falls 20m, in the first 3 seconds it falls 45m and so on. Note that:

$$5 = 5 \times 1^2, \ 20 = 5 \times 2^2, \ 45 = 5 \times 3^2$$

In general, in the first n seconds it falls $5n^2$ metres.

Notice also that no mention is made of the mass of the body. The rate at which the body falls is unaffected by its mass. This fact was well known long before Galileo. He also showed that when a body is thrown at an angle, its path is a parabola. This result follows from the rule above.

Galileo was also a scientist and an inventor. His discoveries include:

- The first thermometer.
- The use of the telescope to discover the moons of Jupiter.
- The use of the telescope to show that the Moon has mountains and valleys (previously it had been thought to be perfectly smooth).
- That a pendulum will swing to and fro in the same time, regardless of the angle through which it swings.

He was a leading supporter of Copernicus's Sun-centered system. This theory was opposed by the Catholic Church and Galileo spent the last years of his life under house arrest.

The Sun-Centered Universe Again

Johannes Kepler (1571–1630)

Three laws about the motion of planets.

The first and most revolutionary law says that planets travel in ellipses rather than circles.

Until Johannes Kepler all heavenly motion was thought to be in terms of circles. The systems of Ptolemy and Copernicus (1473–1543) all had planets moving in circles. Very complicated adjustments involving circles within circles, and off-center circles, had to be made to the theory to make it fit the facts.

By changing circles to ellipses the first law simplified things enormously. There was only one mathematical object, the ellipse, and the Sun was at its focus.

The elliptical model fitted the observed facts very closely indeed.

The second law is that a planet sweeps out equal areas in equal times. So if you draw a line *PS* from the planet (P) to the Sun (S), that line will pass through a fixed area in a fixed time.

The third law is that the cube of the distance of a planet from the Sun is proportional to the square of the length of its year. For Earth the distance is 93,205678 miles and the year is 365 days. For Venus the distance is 67,108088 miles and the Venusian year is 225 days.

Notice that $\dfrac{150^3}{365^2}$ is close to $\dfrac{108^3}{225^2}$.

This law enables us to find the distance of a planet from the Sun, given the length of its year.

All three laws were vital to the development of astronomy. In particular, when Sir Isaac Newton proposed his laws of motion and of gravity, he was able to show that the three Kepler laws could be deduced from them.

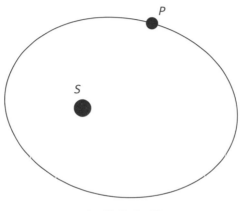

An elliptical orbit.

See: *The Earth-Centered Universe,* pages 48–49; *The Sun-Centered Universe*, page 60; *Three Laws of Motion*, page 90; *The Law of Gravity*, page 91

Logarithms

John Napier (1550–1617)

These are a calculating aid to simplify arithmetic.

Logarithms reduce multiplication to addition and division to subtraction. They were indispensable for arithmetic, especially the multiplication and division of numbers with more than one digit, when this was done by hand before the invention of calculators.

John Napier describes the importance of his invention of logarithms as follows:

Seeing there is nothing ... that is so troublesome to mathematical practice, nor that doth more molest and hinder calculators, than the multiplications, divisions, square, and cubical extractions of great numbers, which besides the tedious expense of time are for the most part subject to many slippery errors ...

The point about logarithms is that when numbers are multiplied, their logarithms are added and when they are divided, the logarithms are subtracted. By hand, addition is much simpler than multiplication. The following is the process of multiplying two numbers using logarithms:

Look up their logarithms in tables.
Add the two logarithms.
Look up in antilogarithm tables.

For example, the product of 2.152 and 3.284 is calculated below using logarithms. The final answer is 7.066.

number	log
2.152	0.3328
3.284	0.5164
7.066	0.8492

But if you use a calculator, the answer becomes:

$2.152 \times 3.284 = 7.067$

This indicates that the result using logarithms is slightly incorrect.

Napier was the Laird of Merchiston in Scotland. The English mathematician Henry Briggs (1561–1630), published Napier's logarithms as tables and helped them gain acceptance among the scientific and academic communities. As simplified by Briggs, in Napier's logarithms each number is converted to a power of 10.

This means that:

- the logarithm of 100 is two (as $10^2 = 2$)
- the logarithm of 1000 is three (as $10^3 = 1000$) and so on.

The logarithms of numbers between integer powers of 10 can also be found.

Until the arrival of affordable calculators in the 1970s, every student of math or science had to be expert at logarithms and they are still used in many African countries. A calculating device based on logarithms, called a *slide rule*, was also common.

Though they are no longer widely used for calculation, logarithms are still important in many parts of mathematics and in science, too, as many quantities are measured by a logarithmic scale.

One example is the intensity of sound which is measured in decibels, in terms of a logarithm of air pressures.

Tessellations

Johannes Kepler (1571–1630)
Roger Penrose (1931–)

A tessellation occurs when a shape is repeated again and again, covering a plane with geometric shapes without leaving any gaps.

The word "tessellate" is derived from the Greek word *tesseres,* which means "four." Tessellation is also known as "tiling."

The question is: What shapes can be used? You can cover a plane with squares and you can also use equilateral triangles or hexagons. The German mathematician Johannes Kepler showed that these are the only regular polygons for which this is possible.

The semi-regular tessellations use two or more regular polygons with the same pattern around each vertex. There are eight of these, one of which is shown. It contains squares and octagons, and is a common pattern on bathroom floors. All of those tessellations are periodic. That is,

they repeat themselves at regular intervals. At every place of the diagram it looks exactly the same.

The physicist Roger Penrose invented several tessellations, built up from two quadrilaterals, known as "kites" and "darts."

A kite   A dart

These tessellations are not periodic: they do not repeat themselves and any two parts of the tessellation are essentially different from each other.

In 1997 the Kimberley Clark company was sued for allegedly using, without permission, Penrose tessellations on its rolls of toilet paper!

Regular tessellation of squares.

Regular tessellation of triangles (left) and hexagons (right).

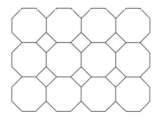

Semi-regular tessellation of squares and octagons.

Regular Solids Revisited

Johannes Kepler (1571–1630)
Louis Poinsot (1777–1859)

Kepler and Poinsot added four new regular solids to the five Platonic solids.

The Platonic solids are all convex. If non-convex shapes are allowed, it means that there are four more regular solids.

A regular solid has faces which are equal regular polygons. The five Platonic solids are all convex; in other words, they do not have recesses or bits sticking out. If we allow the solids to be concave then there are four more solids which fit the definition of regular.

Recall from pages 22–23 that a dodecahedron has 12 faces, each of which is a regular pentagon. Put a five sided pyramid on each face, and we have the Small Stellated Dodecahedron. Another process gives the Great Stellated Dodecahedron.

These solids were discovered mathematically by Kepler. Note the word "mathematically" – artists had found them earlier. A mosaic of the Small Stellated Dodecahedron appears on the floor of St Mark's cathedral in Venice. It is usually attributed to Paolo Uccello (1397–1475). A diagram of the Great Stellated Dodecahedron appears in a 1568 book of geometrical drawings by Wenzel Jamnitzer (1508–1585).

These solids are difficult to understand fully – for example, the centers of the faces are inside the solid rather than on the surface. Still more complicated are the Great Dodecahedron and the Great Icosahedron, found by Poinsot, who also rediscovered Kepler's two solids.

And that is it. That these are the only regular solids was proved by Augustin Cauchy (1789–1857) three years later.

The Platonic solids were important to Kepler. He spent years of observation and calculation trying to show that the solar system fitted precisely into a nest of Platonic solids. Mercury fits into an octahedron, Venus into an icosahedron, and so on. This famous diagram (*right*) of 1596 shows the result.

In terms of modern science it may seem bizarre to model the universe on geometrical shapes. However, this is no more so than the modern practice of modeling the universe on algebraic equations!

Kepler's Platonic solid model of the Solar System from **Mysterium Cosmographicum** *(1596).*

Calculating Machines

Wilhelm Schickard (1592–1635)
Blaise Pascal (1623–1662)
Gottfried Leibnitz (1646–1716)
Charles Xavier Thomas de Colmar (1785–1870)

Mechanical aids for calculation existed even before the invention of calculators and computers.

Although early mechanical calculators were unreliable, they improved over time and were still used until a few decades ago.

As science and commerce became more complicated, efficient calculating methods became crucial. The logarithms of Napier and Briggs were paper-based, but also invented at about the same time were machines for calculation that had dials showing the digits of a number.

There were, however, technical problems with these, particularly with the "carry." If 1 is added to 999,999, it becomes 1,000,000. Because of this, seven dials on the machine have to be moved with one single operation

and early machines were liable to jam at this point.

In the 17th century, German astrologer Wilhelm Schickard designed a machine that could perform the four operations of +, −, x, and ÷. Unfortunately, before it was completed, it was destroyed in a fire, but sufficient details survived for the machine to be successfully reconstructed in the 1950s.

At just 21 years of age, French mathematician Blaise Pascal designed a machine that could add and subtract. It was efficient enough to be commercially produced, under the name of a Pascaline.

German mathematician Gottfried Leibnitz's machine could perform all four operations and he was therefore rather scornful of the limited functions of the Pascaline. Unfortunately, the carry mechanism was faulty and, hence, the machine was never produced in quantity.

The first really successful commercial machine, called an arithmometer, was produced by another Frenchman, Charles Xavier Thomas de Colmar. Built while de Colmar was serving in the French Army in 1820, the machine could perform fairly basic division, addition, subtraction, and multiplication. By the 1860s, it was commercially successful. Under different guises it continued to be produced for well over 100 years.

Mechanical calculating machines were still in use in the worlds of business and education until calculators arrived in the 1970s.

See: *Logarithms,* pages 66–67.

Analytic Geometry

René Descartes (1596–1650)

The invention of analytic (or Cartesian, or coordinate) geometry enabled lines, circles, and so on, to be analyzed by equations.

Analytic geometry is important as it establishes a relationship between geometric curves and algebraic equations.

On graph paper, we can represent the position of a point by two coordinates, the x and the y coordinates. In the diagram P is at the position $(3, 4)$.

We can represent a whole collection of points, such as a line or a curve, by an equation which is obeyed by every point of the collection. The sloping straight line in the diagram contains $(3, 4)$, and also $(2, 3)$, $(1, 2)$, $(0, 1)$, and so on. For all of these points, the y coordinate is always one greater than the x coordinate. Hence, on every point of the line, $y = x + 1$. This is the equation of the line – it defines what the line is.

For circles and other shapes, as well as for lines, we can find an algebraic equation obeyed by every point on the shape. Thus, we can study geometrical shapes by looking at their equations. The intersection of two shapes can be found by solving simultaneous equations. This may not be quicker than proceeding by pure geometry but it is more methodical and reliable. Nowadays there is an extra advantage – there are mathematical computer programs into which we can type equations and get the program to solve them. No computer program can look at a diagram of lines and produce a theorem from it.

There were Greek and Arab precedents for analytic geometry, but it was René Descartes, one of the most important western philosophers, and also a physicist, physiologist, and mathematician, who developed the idea systematically. To some extent

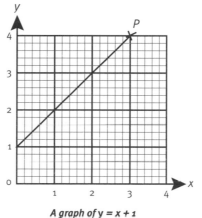

A graph of y = x + 1

modern philosophy begins with his *"Cogito ergo sum"* – *"I think therefore I am."*

Descartes used to lie in bed until midday, a habit which had to change when he became the tutor of Queen Christina of Sweden (1626–1689). She insisted on having mathematics lessons at 5AM in an unheated room, a practice that some experts believe contributed to Descartes's death from pneumonia in 1650.

A Formula for Prime Numbers

Marin Mersenne (1588–1648)
Pierre de Fermat (1601–1665)

The search for a formula for prime numbers has been a long and complicated one.

The formulae $2^n - 1$, where n itself is prime, and $2^{2^n} + 1$ were originally suggested to generate prime numbers.

Euclid (c. 325–265 BC) showed that there are infinitely many prime numbers, but he did not find a rule for generating a succession of them.

The following are all Mersenne primes: $2^2 - 1 = 3$, $2^3 - 1 = 7$, $2^5 - 1 = 31$ and $2^7 - 1 = 127$.

Named after the French priest Marin Mersenne, this type of prime is, in general, a number of the form $2^n - 1$, where n itself is prime. Mersenne thought that all such numbers are prime. This would give us a rule for finding infinitely many primes.

Certainly $2^n - 1$ is prime for $n = 2, 3, 5$, and 7, but the next case, $2^{11} - 1$, turns out to be 23 x 89, and hence not prime.

GIMPS (Great Internet Mersenne Prime Search) is a project for finding larger primes. At the date of writing the largest prime is $2^{32,582,675} - 1$, which has almost 10 million digits and was discovered by Curtis Cooper and Steven Boone in 2006.

Fermat considered the following: $2^{2^0} + 1 = 3$, $2^{2^1} + 1 = 5$, $2^{2^2} + 1 = 17$, $2^{2^3} + 1 = 257$, and $2^{2^4} + 1 = 65,537$.

All these are prime. In general, a Fermat prime is of the form $2^{2^n} + 1$. He conjectured that all such numbers are prime. This is true for $n = 0, 1, 2, 3$, and 4.

All these are prime, but the next case, $2^{2^5} + 1$, turns out to be 641 x 7,600,417, and hence not prime. To date, the only Fermat primes discovered are those listed above: 3, 5, 17, 257, and 65,537.

A formula to generate primes was discovered later but it does not provide a practical method of listing primes.

The Problem of the Points

Pierre de Fermat (1601–1665)
Blaise Pascal (1623–1662)

*The theory of probability began with
this gambling problem.*

**Suppose a game between two players
is interrupted. How should the stakes
be divided? Correspondence between
Fermat, a lawyer and government
official, and the mathematician Pascal
solved this problem.**

Consider a gambling game between
two players in which at each stage one
of the player gains a point. It might be
something as simple as spinning a coin:

if it is Heads, Player A gets a point,

if it is Tails, Player B gets a point.

The game is continued until one of the
players reaches a fixed number of points
and wins, taking all the stakes. It might
be "first to 10" in which case Player A
will win if the coin gives 10 Heads before
it gives 10 Tails.

Suppose this game is interrupted,
however. How should the stakes be
divided? This is the "problem of the
points." Obviously the player who is
ahead should get more than the one
behind, but in what ratio? One early
solution was to divide the stakes in
the ratio of the points already gained.
This is very unfair if the score is 1–0
because the player ahead will get all
the stakes even though he has only a
slender lead.

The Chevalier de Méré, a French
aristocratic gambler, wrote to Pascal
about the problem of the points. A
correspondence then followed between
Pascal and Fermat to discuss the
problem. Fermat's solution, refined by
Pascal, was to look to the future rather
than the past. Instead of considering the
points that had been played, they looked
at the points that would be played if the
game were allowed to continue. The

stakes should be divided in the ratio of their probabilities of winning a continued game.

Consider the coin-spinning game mentioned at the beginning of this entry. Let us assume that two players, A and B, agree that the winner is the "first to 10." Suppose they reach a point in that game at which Player A has 9 Heads (H) and Player B 8 Tails (T). There are up to 2 more spins left in the game. The possible results are as follows:

HH HT TH TT

In three of these results Player A wins because 10 Heads are reached before 10 Tails. Only in the outcome TT (Tails, Tails) does player B win. Therefore, the stakes should be divided in the ratio 3:1, so that Player A gets $\frac{3}{4}$ of the stakes.

The theory of probability began with Fermat and Pascal's work in this field.

Pascal's Triangle

Blaise Pascal (1623–1662)

This is a triangle of numbers designed to find probabilities.

The concept behind Pascal's triangle was known for many centuries before Pascal himself lived. He was the first to use it to calculate probabilities.

Pascal's triangle is the shape of numbers below. Each number is the sum of the two numbers immediately above it.

The triangle was well known in many places and for hundreds of years before the 17th century. In China it is known as Yanghui's triangle, in Iran as Khayyam's triangle and in Italy as Tartaglia's triangle.

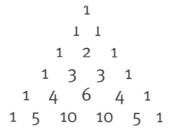

```
          1
        1   1
      1   2   1
    1   3   3   1
  1   4   6   4   1
1   5  10  10   5   1
```

Pascal's triangle: So the next row will be 1, 6, 15, 20, 15, 6, 1.

Before Pascal, uses of the triangle were all algebraic. Below are the expansions of $(1 + x)$, $(1 + x)^2$, $(1 + x)^3$, and $(1 + x)^4$.

$$(1 + x) = 1 + x$$
$$(1 + x)^2 = 1 + 2x + x^2$$
$$(1 + x)^3 = 1 + 3x + 3x^2 + x^3$$
$$(1 + x)^4 = 1 + 4x + 6x^2 + 4x^3 + x^4$$

Ignore the x terms. Just look at the coefficients:

```
1   1
1   2   1
1   3   3   1
1   4   6   4   1
```

These are rows from Pascal's triangle. The next row is 1, 5, 10, 10, 5, 1. So, to expand $(1 + x)^5$, we don't need to do any heavy algebra. Just write down the result, using the next row of the triangle:

$$(1 + x)^5 = 1 + 5x + 10x^2 + 10x^3 + 5x^4 + x^5.$$

To be fair to Pascal, he did not claim to have invented the triangle. He certainly did not name it after himself. He actually referred to it as the "arithmetic triangle" and used it to calculate probabilities. The next topic, *The Binomial Distribution*, shows how it was used.

The Binomial Distribution

Pierre de Fermat (1601–1665)
Blaise Pascal (1623–1662)

Binomial distribution is a probability distribution for the number of successes in a series of experiments.

Suppose a die is rolled several times. The binomial distribution gives the probability of obtaining a certain number of sixes.

Spin a coin once. There are two possible results, Heads (H) and Tails (T). Spin the coin twice, and there are four results:

HH HT TH TT.

Two of these four results give exactly one Head (HT and TH) so the probability of exactly one Head is $^2/_4 = {}^1/_2$. With three spins, the eight results are:

HHH HHT HTH HTT
THH THT TTH TTT.

The probability of exactly one head is $^3/_8$, and so on.

With four spins, there are 16 outcomes and with five spins there are 32. It would be very tedious to write out all these 32 outcomes. Here Pascal's triangle comes to our aid. Row 5 of the triangle contains the numbers 1, 5, 10, 10, 5 and 1. With five spins there could be 0, 1, 2, 3, 4 or 5 Heads. The probabilities of these are found by taking the row of numbers 1, 5, 10, 10, 5, 1 respectively, and dividing each by 32. Therefore, the probability of exactly 2 Heads in five spins is:

$$^{10}/_{32} = {}^5/_{16}.$$

In general, the binomial distribution enables us to find the probabilities without having to list the results. Suppose an experiment has probability *p* of success (such as a Head for a coin, or a 6 for a die) and that the experiment is repeated independently n times. Then the probability of exactly r successes is:

$$^{n}Cr\, p^{r}(1-p)^{n-r}.$$

Here ^{n}Cr is the r term from the n row of Pascal's triangle. It can be found directly from a scientific calculator.

The binomial distribution is relevant to many games of chance. It is also central to much of statistics. Take opinion polling for example. Suppose that in a certain state of America the proportion of Republican voters is 40 per cent (0.4 in other words). An opinion poll of 1,000 voters is taken. Hence, we can experiment by asking a voter for whom he or she intends to vote. The probability of the voter intending to vote Republican is 0.4. This experiment is repeated 1,000 times. The binomial distribution gives the probabilities of the numbers of Republican voters in the sample of 1,000. For example, the probability of 450 out of the 1,000 people saying that they would vote for the Republicans is actually:

$$^{1000}C_{450}(0.4)^{450}(0.6)^{550}.$$

From these probabilities we can predict the result of a future election in the state.

Pascal's Wager

Blaise Pascal (1623–1662)

*Pascal enlisted the theory of probability
to support religious belief.*

The mathematician Pascal contributed towards many fields. His co-invention of probability has already been mentioned. His investigations into the atmosphere and the vacuum were also important, so much so that the unit of pressure, the pascal, is named after him.

Pascal is also famous for the *Pensées*, a collection of notes which formed the basis for an intended book justifying Christianity. He died before he was able to write the book itself. The notes, published posthumously in 1670, contain the following, known as the Wager:

God is, or he is not. At the far end of an infinite distance a coin is being spun which will come down Heads or Tails. How will you wager?

You must wager. It is not optional. There is an infinitely happy life to gain, and a chance of gain against a finite number of chances of loss, and what you stake is finite. It is no use saying that whether we gain is uncertain, while what we risk is certain...

The argument here is that the reward for believing in God is infinite while what we surrender for that belief, a few worldly pleasures, is finite. However small the probability of God's existence, the gain from belief is greater, infinitely greater, than that from disbelief.

Differentiation

Sir Isaac Newton (1642–1727)
Gottfried Leibnitz (1646–1716)

Differentiation can be used to find the gradient of a curve.

———

The gradient of a straight line is simple to calculate. The gradient of a curve is harder to calculate as it is constantly changing.

To find the gradient or slope of a straight line, divide the "rise" by the "run" (*see left-hand diagram on page 85*). For a road, this is the vertical distance risen, divided by the horizontal distance traveled.

If the line is not straight, the gradient is constantly changing. In the right-hand diagram, the curve starts off flat, with zero gradient, then gets steeper and steeper.

At no section of the curve is there a constant gradient. The best we can do, to find the gradient at a point, is to find the gradient of the tangent at that point.

Sir Isaac Newton and Gottfrieb Leibnitz, working in the United Kingdom and Germany, found the gradient of the tangent by taking an infinitely small rise divided by an infinitely small run. The subject they founded was infinitesimal calculus, and this branch of it – finding the gradients of curves – is differentiation.

Following this discovery, however, there was a long and unpleasant dispute about who could claim the credit, which lasted about 25 years. It seems that Newton got there first but he kept it secret. Leibnitz discovered it independently a decade or so later and was the first to publish the results.

Suppose the curved graph (*see the right-hand diagram on page 85*) represents the distance a car has traveled. The gradient then represents the rate of change of the distance (or the speed) of the car. In this case, the car starts off at rest and gradually accelerates. At no time does the curve have a constant gradient, so

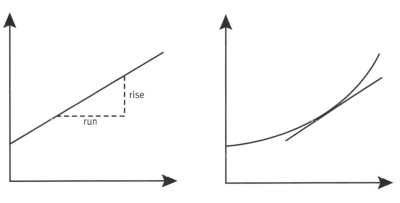

Gradient of a straight line.

Gradient of a curve.

at no time does the car have a constant speed. The gradient found by differentiation represents the instantaneous speed, which is what is shown on the speedometer.

Differentiation can therefore be used to find speed and acceleration and many other physical quantities. Calculus was, and still is, the most powerful tool in science.

Integration

Sir Isaac Newton (1642–1727)
Gottfried Leibnitz (1646–1716)

Integration concerns finding the area under a curve.

If a shape has straight edges its area can be found by cutting it into triangles. Finding the area of a shape with curved edges is trickier.

For some shapes with curved edges the area had been measured; the area of a circle was known, of course, and Archimedes had found the area between a parabola and a straight line. A general method for finding the area of a general shape with curved edges had not been found until the invention of integration.

Suppose the shape is as shown. Newton and Leibnitz found the area by dividing it into infinitely many infinitely thin vertical strips, and adding their areas. This was essentially the same method as used by Archimedes.

Gottfried Leibnitz and Sir Isaac Newton developed calculus independently of each other and each devised his own notation. For both differentiation and integration the symbols used by Leibnitz are more generally used than Newton's. The infinitesimal changes in x and y are written as dx and dy.

Differentiation is shown as dy/dx, the dy and dx being the infinitely small rise and run.

Integration is shown as $\int y dx$. The infinitely thin strips in the diagram are each a rectangle with height y and width dx, and hence with area $y dx$. The $\int$ sign is an old-fashioned "s" which stands for "summa," indicating that these areas should be summed, or added.

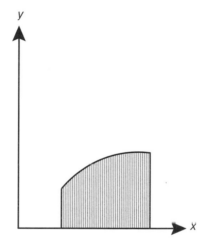

Area under a curve.

See: *Differentiation,* pages 84–85

The Fundamental Theorem of Calculus

Sir Isaac Newton (1642–1727)
Gottfried Leibnitz (1646–1716)

This theorem shows the relationship between differentiation and integration, the two main operations of calculus.

If a function is integrated and then differentiated, it is back to the original function.

Addition and subtraction are inverse procedures. Think of a number, add 7, subtract 7, and you are back to the original number. Multiplying and factorizing are inverse operations. Multiply 17 by 11 and you get 187. Factorize 187 and you get back to 17 and 11.

On a more advanced level, integration and differentiation are inverse to each other. Think of an expression in *x*, integrate it, differentiate the result, and you are back to the original expression.

This is known as the fundamental theorem of calculus. It joins the two departments of calculus. The infinitesimal calculus comprises differentiation, integration, and the fundamental theorem linking the two. Calculus was spectacularly fruitful, both within mathematics and in its application to science. It gave correct results even though its foundations were very dubious.

Both differentiation and integration involve infinitesimals which Sir Isaac Newton called "fluxions." He kept them a secret, perhaps because the idea of a number that was at the same time infinitely small and yet non-zero went against all the principles of arithmetic.

In differentiation, how can one divide one infinitely small number by another and obtain a sensible result?

In integration, how can the sum of infinitely many things of zero area be something non-zero?

The whole logical foundation of calculus was suspect and was attacked by philosophers of the time. George Berkeley (1685–1753), Bishop of Cloyne, commented in his discourse *The Analyst, addressed to an infidel mathematician*:

And what are these Fluxions? ... They are neither finite Quantities nor Quantities infinitely small, nor yet nothing. May we not call them the Ghosts of departed Quantities?

By 1734, the time at which Berkeley was writing, the "infidel" in question, Newton, had been dead for 7 years. and his invention of calculus was far too successful to be abandoned. It was put on a logically sound foundation in the 19th century.

See: *Differentiation,* pages 84–85; *Integration,* pages 86–87

Three Laws of Motion

Sir Isaac Newton (1642–1727)

*Sir Isaac Newton developed the three laws of
motion in the 1660s, but they were only
made public in the late 1680s.*

These laws occur at the beginning of *Philosophiae Naturalis Principia Mathematica*, (*Mathematical Principles of Natural Philosophy*), a book with a claim to be the greatest ever written.

These laws were unquestioned for over 200 years and still hold in all but the most extreme of circumstances. They state:

1. *A body remains at rest or in motion in a straight line unless a force is impressed on it.*

Thus force is needed to *stop* motion or to change its direction, as well as to *start* motion.

2. *The change in motion is proportional to the force, and is in the same direction as the force.*

Thus the *acceleration* of a body is proportional to the force and is in the same direction.

3. *To every action there is an equal and opposite reaction.*

Thus, if body A exerts a force on body B, body B exerts an equal and opposite force on A.

The Scientific Revolution is said to have begun with Copernicus and it reached its climax with Newton, whose influence on science and ideas has been much praised ever since.

The poet William Wordsworth wrote of him:

*...a mind forever,
Voyaging through strange seas of
thought, alone.*

The Law of Gravity

Sir Isaac Newton (1642–1727)

Any two bodies in the universe attract each other, according to the law of gravity. This is a law that applies to every object in the universe.

The most famous story in science is that of Sir Isaac Newton and the apple: it fell from a tree in his garden and inspired the theory of gravity.

Two things could be deduced from this everyday occurrence, things which are obvious today but were revolutionary at the time:

- The apple fell because there was a force pulling it.
- The force pulling the apple was the same force that held the Earth in orbit around the Sun.

The law of gravity, which applies equally to falling apples and to planets and stars, is that bodies of mass m and M, a distance d apart, attract each other with a force of GmM/d^2. Here G is the universal constant of gravity, about $0.000,000,000,067 \text{ Nm}^2\text{kg}^{-2}$.

One big question remained. How is this force transmitted across the space between the Sun and the Earth, where there is nothing to transmit it?

Newton fudged the issue, saying: *"Hypotheses non fingo,"* that is, *"I don't deal in hypotheses."*

The Precession of the Equinoxes

Sir Isaac Newton (1642–1727)

*The position of the Sun relative to the stars
at the spring equinox is not constant.*

The year defined by the seasons is slightly different from the year defined by the stars. Sir Isaac Newton explained this in *Principia Mathematica*.

Suppose you were born on 21 March, the spring equinox. According to astrology, your star sign would be Aries the Ram, because on 21 March the Sun is in the constellation of Aries.

In reality, this is not the case, however. On 21 March 2009 the Sun is in the constellation of Pisces the Fish, one star sign on from Aries. The astrological calendar is out of phase with the ordinary calendar. The year between successive equinoxes is slightly different from the year defined by the position of the Sun against the stars, by about 20 minutes.

This was first noticed by Hipparchus in the 2nd century BC. The system of astrological star signs had been established when the Sun had been in Aries at the time of the spring equinox. That was many centuries earlier and the position of the Sun had moved since. This phenomenon is the precession of the equinoxes: the position of the Sun relative to the stars changes, making a full circle round the Zodiac every 25,765 years.

Hipparchus could give no reason for it, but it was explained by Newton centuries later. The Earth's radius is greater at the equator than at the poles, and the difference in gravitational pull by the Sun causes the axis of the Earth's rotation itself to rotate, rather like the axis of a spinning top.

The Law of Large Numbers

Jacob Bernoulli (1654–1705)

Suppose an experiment has a probability p of success. With more and more trials, the proportion of successes approaches p, with probability one. Suppose a fair coin is spun repeatedly. Almost certainly the proportion of Heads will approach 1/2.

The law of large numbers, as formulated by Swiss mathematician Jacob Bernoulli, appeared in his manuscript *Ars Conjectandi*, which was published after his death in 1705. The Bernoulli family contained eight mathematicians, all of whom made significant contributions to mathematics. This Jacob is Jacob the First; Jacob the Second was his great nephew.

When asked for a definition of the "Law of Averages," the philosopher C. E. M Joad (1891–1953) replied: "*If you spin a coin a hundred times, it will come down heads 50 times, and tails 50 times.*"

The learned professor should have known better. The probability of exactly 50 heads out of a hundred spins is only about 0.08. The law of large numbers, which is the mathematical phrase for the law of averages, says that the proportion of spins that give Heads will *approach* $1/2$, not that *exactly* half of the spins will give Heads.

Moreover, we are not absolutely certain that this will happen. It is physically possible for the coin to come up heads for all eternity. The probability of this happening is zero however. Hence the law has to be qualified: that with probability one, the proportion of Heads approaches $1/2$.

The Normal Distribution

Abraham de Moivre (1667–1754)

This is a probability distribution with the familiar bell-shaped curve.

This probability distribution approximates the binomial distribution. It is used to model statistical data from many sources.

Earlier in the book we introduced the concept of the binomial distribution. If the number of experiments is large, then it will be very tedious to work out probabilities. In the example of opinion polling, we might want to know the probability that there are fewer than 450 Republican voters in the sample of a 1000. This would involve working out 450 separate probabilities!

What Abraham de Moivre showed was that, provided that the number n of trials is large, the binomial distribution can be accurately approximated by another distribution, called the normal distribution. The graph on the opposite page shows the binomial distribution for the number of Heads when a fair coin has been spun 15 times. The normal distribution is the smooth curve, which fits the binomial distribution, the collection of rectangles. The area in the rectangles is close to the area under the curve.

For the opinion poll problem, instead of finding the area in 450 separate rectangles we can find the area under one single curve.

The normal distribution was used in many other circumstances: to model experimental errors and astronomical data. It can be used to model things that cluster round an average, such as the heights of adult women, examination scores, temperatures, or many other observations.

The normal curve has many other names: the error curve, the Gaussian curve, and the bell curve.

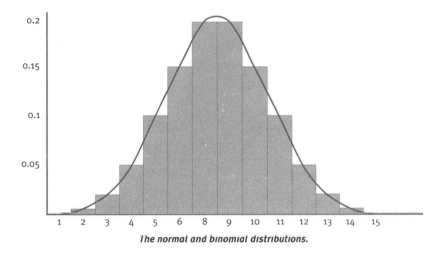

The normal and binomial distributions.

This last name achieved notoriety in 1994 as the title of a book by Richard Herrnstein and Charles Murray, which was accused of lending support to the view that certain races are inherently intellectually inferior to others, for which the authors received heavy criticism.

See: *The Binomial Distribution,* pages 81–82.

The Seven Bridges of Königsberg

Leonhard Euler (1707–1783)

The city of Königsberg had seven bridges. Was it possible to walk around the city crossing each bridge exactly once? This was a popular puzzle that engendered a whole branch of mathematics.

In the 18th century the East Prussian city of Königsberg (now Kaliningrad, in Russia) had seven bridges crossing the river Preger. A simplified map of the city is shown below.

Was it possible to tour the city, crossing each bridge once and once only? This was a popular puzzle of the time.

If a route could be found, it would have been found quickly. The mathematics to show that no such route is possible did not exist at the time until, Euler invented it.

Change the map, so that each region of land is represented by a dot and each

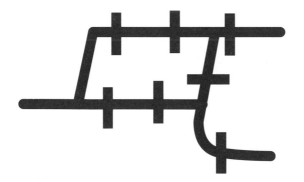

A simple map of Königsberg.

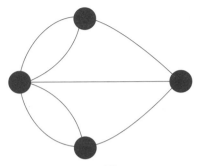

Dots and lines.

bridge is represented by a line joining the dots *(above)*. The leftmost dot represents the central island, which has five bridges leading to it.

The problem now becomes to draw the diagram without lifting pen from paper and without going over any line twice.

The key to the solution is to count the number of lines leaving each dot. At the beginning of the journey one can leave a dot without returning to it. At the end of the journey one can reach a dot without leaving it. But consider a middle stage of the journey. Every time one reaches a dot one must then leave it. However many times the dot is reached, it must also be left.

So the number of lines from such a middle dot must be even. At most, two of the dots (the beginning of the journey and its end) can have an odd number of lines from it.

In the Königsberg diagram all four of the dots have an odd number of lines leaving from them. Therefore, it is impossible to draw the diagram continuously and without retracing a line, and so it is impossible to walk around Königsberg crossing each bridge once and once only.

Euler's solution became the basis for the foundation of graph theory and, indirectly, of combinatorics and topology.

Goldbach's Conjecture

Christian Goldbach (1690–1764)

Every even number above two is the sum of two primes.

Although the Goldbach conjecture **has been verified up to a very large number, general proof is still elusive.**

Take the first few even numbers after 2. We can write each of them in terms of two primes added together. You are allowed to repeat a prime:

4 = 2+2   6 = 3+3   8 = 3+5
10 = 3+7   12 = 5+7   and so on.

Every even number that has been tested has been shown to be the sum of two primes. Goldbach was certain of the truth of his conjecture though he admitted he was unable to prove it.

The conjecture is true up to 10^{18}, which is one followed by 18 zeros, that is:

1,000,000,000,000,000,000.

Even if there is a counterexample (an even number that *cannot* be written as the sum of two primes) then it must be very large. It must be greater than 10^{18}.

The conjecture refers to *two* primes. It has been shown that every even number is the sum of a certain number of primes, but so far the number is greater than two. It was shown in 1939 that every even number is the sum of 300,000 primes. This is quite a long way from two! This was whittled away, and in 1995 it was down to six primes.

The conjecture is sometimes known as the "strong Goldbach conjecture." The weak conjecture is that every odd number after 7 is the sum of three primes. It is also unproven, so it is not very weak!

V + F = E + 2

Leonhard Euler (1707–1783)

This is a formula connecting the vertices, faces and edges of a solid shape.

A polyhedron has flat faces and straight edges. This simple formula works for polyhedra which are convex.

The face of a solid is a flat shape on its surface. Two faces meet at an edge. Three or more edges meet at a vertex.

The five Platonic solids are polyhedra. The table below gives the number of vertices (V), faces (F), and edges (E) for each shape.

SOLID	V	F	E
Tetrahedron	4	4	6
Cube	8	6	12
Octahedron	6	8	12
Dodecahedron	20	12	30
Icosahedron	12	20	30

For example, for the cube shown, the point A is a vertex, the square ABCD is a face, the line AB is an edge. The cube has 8 vertices (4 on the top and 4 on the bottom), 6 faces (2 horizontal and 4 vertical) and 12 edges (8 horizontal and 4 vertical).

These solids had been studied extensively but, before Euler, no one had noticed the relation connecting V, F, and E. For each row, add the V and F numbers and the result is always 2 greater than the E number.

The formula holds for convex polyhedra: a solid is convex if a line joining any two points of the solid lies entirely within the solid. It does not hold for the two Kepler solids. In general, it does not hold for solids with holes in them.

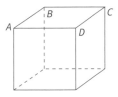

The Gambler's Fallacy

Jean d'Alembert (1717–1783)

This is the belief that past occurrences can influence the future.

A fair coin is spun and comes up Heads 10 times. The gambler's fallacy is that the probability of a Head at the next spin is less than $\frac{1}{2}$.

Two events are independent if the result of one does not influence the probability of the other. So if a fair coin is spun and gives Heads, the probability that the next coin also gives Heads is still $\frac{1}{2}$. The French mathematician Jean d'Alembert had a rare lapse when he denied this in his article on probability in the *Encyclopédie* (a great work covering all human knowledge at the time).

To extend this concept, however many times in a row a fair coin gives Heads, the probability of a Head next time is still $\frac{1}{2}$. The gambler's fallacy denies this. It maintains that after a run of Heads, a Tails throw is more likely.

The gambler's fallacy is essentially a magical belief and relies upon the great slogan of magic, from an ancient text of alchemy, which is: *As above, so below*.

Hence, if there is a certain configuration of the stars and planets, that must be replicated in the lives of humans on Earth. Astrology is based on this belief.

The theoretical distribution of Heads and Tails in spins of a fair coin is that they should be equal. The gambler's fallacy is that this theoretical distribution should be precisely modeled in reality. This belief has certainly benefited casinos: they have grown rich on it.

Complex Numbers

A complex number is of the form a + *i*b,
where a *and* b *are real and* $i^2 = -1$.

The square of any ordinary number is positive. So numbers whose square is negative have to be invented.

The square root of −1, or √−1, is something that does not exist in ordinary numbers. Any ordinary number, positive or negative, becomes positive when squared. If you try to find √−1 on a calculator, you will get an error message.

There are many equations which involve the square root of a negative number. The simplest of these is: $x^2 = -1$. The obvious way to deal with such an equation is to say that there is no solution. Throughout the 16th and 17th centuries it became apparent that it would be very convenient if solutions did exist. The method of solving the cubic equation sometimes involved the square root of a negative number, even for the simple equation: $x^3 - x = 0$.

This struck Tartaglia as rather absurd, especially as the solutions of this equation are the simple ordinary numbers 0, 1 and −1.

If ordinary numbers cannot involve the square root of a negative number, it would be useful to have numbers that can and so the answer was to *invent* numbers to fill the gap. The square root of −1, √−1, is written i (though electrical engineers call it j). A complex number is found by mixing i with ordinary numbers, so we get anything of the form $a + ib$.

This sort of invention is not new in mathematics. After all, it makes no sense to say: "*there are minus five people in the room*." It is often convenient to use negative numbers, however, with temperature, bank balances, and so on, and hence they became accepted. Similarly, throughout the 18th century complex numbers became accepted, and by the 19th century, mathematicians could not have carried on without them.

With complex numbers, mathematics is richer, more regular, and more satisfying.

To take a familiar case, a quadratic equation may or may not have solutions with real numbers. With complex numbers it always has two solutions.

Complex numbers are not just for pure, theoretical mathematics. They are used in many fields of science and technology as well. For example, complex numbers are used by electrical engineers to describe the impedance of an electrical circuit, what hinders a current from flowing round it.

See: *Negative Numbers,* pages 46–47

e

Leonhard Euler (1707–1783)

The number e is such that the gradient of $y = e^x$ is equal to itself.

The number e is one of the most important in mathematics. The e stands for "exponential:" an exponential function is one for which its rate of growth is proportional to itself.

The process of differentiation finds the gradient of a curve. There is a function which remains the same when it is differentiated; in other words, it is equal to its own gradient. The function is e^x, where e is a number approximately equal to 2.718. So if $y = e^x$, $\frac{dy}{dx}$ is also equal to e^x.

This may seem to be just a mathematical curiosity but there are many natural phenomena which increase or decrease at a rate proportional to themselves. Examples include:

• Population. Unchecked, a population grows at a rate proportional to itself.

• Radioactivity. A radioactive material decays at a rate proportional to itself.

The number e, and the function e^x, occur throughout applied mathematics and statistics.

The number was given the letter e by Euler, who must have been pleased that it was the first letter of his surname, as well as standing for exponential.

Euler was one of the most prolific mathematicians in history. It is estimated that his complete works would fill 50 fat volumes. One of his many results is a famous equation that links four of the most important numbers in mathematics: e itself, i, π, and –1. The equation is:

$$e^{i\pi} = -1.$$

See: *π,* pages 15–16; *Negative Numbers*, pages 46–47

Regular Polygons Revisited

Carl Friedrich Gauss (1777–1855)

A regular 17-sided figure can be constructed.

Greek mathematicians could construct, using straight edge and compasses only, polygons with 3 and 5 sides. German mathematician Carl Friedrich Gauss devised a method of extending this to a 17-sided polygon.

Recall that for an exact construction one is only allowed a straight edge and compasses. The use of a protractor is not allowed when measuring angles nor a ruler to measure distances.

In 1796, Gauss showed how to construct a regular 17-sided polygon. He was so proud of this result that he wanted the shape to be inscribed on his tombstone, but the stonemason refused.

What is significant about 17? This links up with another topic. The number 17 is a Fermat prime , as it is $2^{2^2} + 1$. What Gauss showed was that a regular n-sided figure can be constructed if n is a Fermat prime.

A regular heptagon (7 sides) cannot be constructed, as the number 7 is not a Fermat prime.

The next Fermat primes are 257 and 65,537. In 1832, F. J. Richelot, who taught mathematical analysis at the University of Königsberg, showed how to construct a 257-sided polygon.

In 1894, J. Hermes wrote out the construction of a regular 65,537-sided figure. It had taken him 10 years and took up 200 pages. Unfortunately it is likely to contain a mistake.

See: *A Formula for Prime Numbers,* page 76; *Regular Polygons*, page 21

Arithmetic and Geometric Progressions

Thomas Malthus (1766–1834)

These are two increasing sequences,
which have applications to population.

Arithmetic and geometric progressions have been known since the time of the Greeks. English-born economist Thomas Malthus used them to describe the growth of population.

A sequence is arithmetic if each term is found by adding a constant to its predecessor.

The following sequence, which goes up in steps of three, is arithmetic:

5   8   11   14   17   ...

A sequence is geometric if each term is found by multiplying its predecessor by a constant.

The following sequence, in which each term is multiplied by two, is geometric:

3   6   12   24   48   ...

A well-known example of a geometric sequence is about money left in the bank at a constant rate of interest. If the interest rate is 5 per cent, then the amount will be multiplied each year by $^{105}/_{100}$, which is 1.05. Suppose the original amount is $1,000. In successive years, the amounts are, to the nearest $:

$1,000 $1,050 $1,103 $1,158
$1,216 $1,276

If the difference between terms in an arithmetic progression is positive, and if the ratio between terms in a geometric progression is greater than one, then both sequences will increase without limit. Both sequences are increasing, but in the long run the geometric progression wins. Any such geometric progression will eventually overtake any such arithmetic progression.

This mathematical fact inspired Malthus. He was the founder of demography, the study of human populations. In Malthus's 1798 "An Essay on the Principle of Population," he wrote that population would increase geometrically, while the increase in food resources would be only arithmetic. The gloomy prediction was that population would outstrip food supply with famine and war as the result.

In Malthus's time Great Britain had a population of about 10 million. This has since grown to over 60 million. Whatever the problems of modern Britain, shortage of food is not among them. Fortunately, therefore, Malthus's dire warning has not been borne out.

The Fundamental Theorem of Algebra

Carl Friedrich Gauss (1777–1855)

An equation of degree n has n roots.

Using ordinary numbers, a quadratic equation has zero, one, or two solutions. If we allow complex numbers, it has exactly two solutions. The fundamental theorem of algebra extends this to all polynomial equations.

Many topics in this book concern the solution of equations. The quadratic equation can have zero, one, or two solutions in ordinary numbers. The graphs below illustrate the three cases. Notice that the graph crosses the x axis zero times, one time or two times.

If we allow complex numbers, then any quadratic equation can be solved. There are always two solutions if we say that a perfect square has one solution that is counted twice, for example, $(x-1)(x-1) = 0$ has solutions 1 and 1.

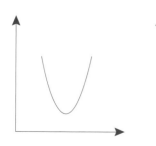

No real solutions.

One real solution.

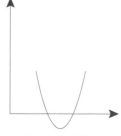

Two real solutions.

The fundamental theorem of algebra extends this to all polynomial equations. If an equation has degree n – in other words, the highest power of x is x^n – then there are n solutions.

The proof is non-constructive: it merely says that these solutions exist, without providing a method to find them. As Abel and Galois later showed, there is no algebraic formula to find the solutions if n is five or more.

The theorem provides an example of where the theory of complex numbers is more complete than that of real numbers.

See: *Quadratic Equations,* pages 11–12; *Complex Numbers*, pages 101–102; *Quintic Equations,* page 112; *Galois Theory*, page 115

Fourier Series

Joseph Fourier (1768–1830)

Any wave form can be made by combining sine and cosine functions.

Introduced as a solution to the heat equation, the Fourier series can be applied mathematically in many different ways, for example, to analyze waves that enable the sound of any musical instrument to be reproduced.

The first graph shows the function: $y = \sin t$. If t represents time and y the air pressure, then this graph represents a sound. The sound would be very pure.

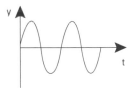

The note produced by a flute follows the pattern of the above diagram. Other instruments produce more complicated sounds with more complicated waves. A bass clarinet would be like this:

As well as the basic note, the sound contains notes with twice the basic frequency, thrice the basic frequency, and so on. These extra notes are called overtones.

Graphs like those here, which repeat themselves after a fixed length of time, are called "periodic." Fourier showed that any periodic function can be expressed as a sum of sine and cosine functions. For example: $\frac{1}{2} \sin t + \frac{1}{4} \sin 2t + \frac{1}{8} \sin 3t$.

This will give the same note as that in the flute curve, but the overtones make the sound richer and more complicated. Music synthesizers can use this to reproduce the sound of any instrument.

The Difference and Analytic Engines

Charles Babbage (1791–1871)

These are machines that performed calculation automatically.

The difference engine could evaluate functions automatically, while the analytic engine was the first programmable computer.

"I am thinking that all these tables might be calculated by steam," said English mathematician Charles Babbage when looking at logarithm tables and contemplating the dreary and error-prone task of compiling them. It sounds quaint today, but in the early 19th century, steam power had made manufacturing and transport more rapid and reliable. Why, Babbage thought, should it not do the same for calculation?

Babbage's first invention was the difference engine. This had a fairly precise task – to calculate and print out the values of certain functions. Known values of the required function were entered and the machine then calculated other values. It would be exact for polynomial functions up to a certain level, and approximate for logarithmic or trigonometric functions. Babbage never finished his version but simplified models were made in about 1855.

In 1990, the Science Museum in London built the machine following Babbage's design and it worked to an accuracy of 31 decimal places.

The analytic engine was a much more ambitious project. In many ways it was equivalent to a modern computer, except that it worked mechanically rather than electronically. It was programmable, in that it could be instructed to carry out any mathematical task. These instructions were entered on punched cards, as were the data on which the instructions would work. It could branch (it could take different directions depending on the result of a calculation) and it could loop (perform an operation over and over again until a certain result was reached).

In all, it was "Turing complete." In theory it could perform any computing task that can be performed. "In theory" because the machine would have taken an unfeasibly long time to perform many of the tasks of modern computers. To multiply together two 20-digit numbers would have taken about 3 minutes, while a modern computer does this in a billionth of a second.

The analytic engine was actually never built. Babbage kept changing his mind about what he wanted and quarreled with everyone working with him. The British government which was meant to provide funding for the project, withdrew from it in 1842, when it seemed unlikely that anything would come of it.

"We got nothing for our £17,000 but Mr Babbage's grumblings," wrote the secretary of the Royal Astronomical Society. *"We should at least have had a clever toy for our money."*

Today, a century-and-a-half later, the descendants of Babbage's analytic engine are much, much more than clever toys.

See: *Turing Machines*, page 171

Quintic Equations

Niels Abel (1802–1829)

Methods of solving quadratic equations have been known for thousands of years, while the cubic and quartic equations were solved in the 16th century. But what about the quintic equation, the equation in which the highest power of x is 5? Can there be a formula, involving fifth roots as well as fourth, cube, and square roots, which gives the solution?

A t each stage a clever trick was found to solve quadratic, cubic, and quartic equations. At first it seems as if a new clever trick must be found for the quintic, then another trick for the sextic (involving x^6) and so on for ever, which can seem daunting. But a surprise result showed that the sequence of formulae for equations ends with the quartic.

There is no general formula for the solution of the quintic:
$$ax^5 + bx^4 + cx^3 + dx^2 + ex + f = 0.$$

Of course, some quintics can be solved. The equation $x^5 - 32 = 0$ has the solution $x = 2$. The equation $x^5 = 2$ has the solution $x = \sqrt[5]{2}$, the 5th root of 2. However, in general, there is no formula involving taking roots that will work in every single case.

Many other mathematicians had tried to solve quintic equations, but Norwegian-born Niels Abel showed it as impossible in 1821. At the time Norway was a poor backwater and, despite his discoveries, Abel was not fluent enough in French or German to be certain of obtaining a well-paid position. He suffered from ill health, exacerbated by poverty. A post was at last found for him in Berlin, Germany, but news of his success did not arrive until a few days after his death.

Non-Euclidean Geometry

Nicolai Lobachevsky (1792–1856)
Janos Bolyai (1802–1860)

These are systems of geometry in which Euclid's fifth postulate no longer holds.

Euclid's fifth postulate is far from obvious and many attempts were made to prove it from the other postulates. Eventually, types of geometry were constructed in which the postulate is false.

Euclid's fifth postulate is also called the parallel postulate. Many important results follow from it, for example that the sum of the angles in a triangle is 180°. There were various attempts to prove the postulate from other axioms, all of which were shown to need some other assumption. The best-known of these assumptions is Playfair's Axiom.

It is equivalent to the fifth postulate (*see diagram*):

Given a line and a point P not on the line, there is exactly one line through the point parallel to the line.

This excludes two possibilities:

1. That there are no parallels through *P*.
2. That there is more than one parallel through *P*.

There are versions of geometry in which either points 1 or 2 are true. To take these in turn:

1. In an elliptic geometry there are no such things as parallel lines. Any two lines must meet somewhere. This is best imagined as the top half of the

A parallel through P

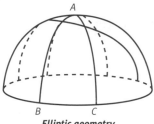

Elliptic geometry

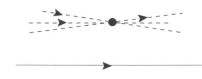

Hyperbolic geometry

sphere, in which lines are represented by half circles whose centers are the center of the sphere.

Notice that any two lines do meet somewhere on the surface, so there are no parallel lines. Also note that in triangle ABC, ∠ABC and ∠ACB are both 90° and ∠BAC is not zero. So the sum of the angles in the triangle is greater than 180°.

2. In hyperbolic geometry there are many lines through a point parallel to a given line. In the diagram, none of the dotted lines ever meet the solid line, and hence all are parallel to it. This can be modeled by a hyperbolic surface which curves inwards rather than outwards. In this geometry, the sum of the angles of a triangle is less than 180°.

The discovery of non-Euclidean geometry fascinated many people besides mathematicians. In the acclaimed novel *The Brothers Karamazov* by Fyodor Dostoyevsky (1821–1881), Ivan tells his brother Alyosha:

> ... if God created the world, he created it in accordance with Euclidean geometry, and yet there are mathematicians, men of extraordinary genius, who doubt whether the whole universe was created only in Euclid's geometry ...

Thus, the discovery of non-Euclidean geometry threw doubt on the very existence of God.

See: *The Fifth Postulate*, pages 36–37; *Sum of the Angles in a Triangle*, page 38

Galois Theory

Evariste Galois (1811–1832)

Galois theory is a branch of mathematics derived from the theory of equations.

Galois theory analyses the solutions of an equation in terms of permutations. It led to group theory, one of the major topics of mathematics.

Many of the topics in this book have been concerned with solving equations such as the quadratic, the cubic, the quartic, and the quintic. The first three can be solved by formulae, the last cannot. Galois theory puts it all together, considering the possible solutions of an equation, and showing how they can be interchanged. These interchanges are called "permutations" and the properties of these permutations decide how the equation can be solved.

The permutations form what Galois called a group. The groups of permutations corresponding to the quadratic, cubic, and quartic have a property that the quintic doesn't, and

that is why there is no general formula for the quintic. Group theory is now one of the major areas of mathematics. It has been extended to many other structures besides permutations and is used In physics, as well as in pure mathematics.

Galois was a tragic hero. His teachers did not have a high opinion of him and he failed exam after exam. He sent off his work to leading mathematicians of the day, but it was so new, and also so badly written, that few recognized its worth.

Galois was arrested for making what was interpreted as a threat to murder the new king, Louis Philippe in 1831. He was acquitted but was arrested again on Bastille Day of that year. This time he was imprisoned. On leaving prison he was killed in a duel. His work was published posthumously in 1846.

Constructible Lengths

Pierre Wantzel (1814–1848)

This provides a description of the lengths that can be constructed.

The three Greek problems of doubling the cube, trisecting the angle, and squaring the circle all involve constructions using straight edge and compasses only. What lengths can be constructed using these instruments?

Many ingenious methods were suggested for solving these three famous problems. They can be solved, but only with instruments beyond those of straight edge and compasses.

The Greeks did not have the mathematics to show that these constructions are not possible. To show this, it was first necessary to identify which lengths *can* be constructed.

With straight edge and compasses, there are essentially five basic operations:

1. Joining two points with the straight edge.

2. Drawing a circle with a given center, through a given point.
3. Finding the intersection of two straight lines.
4. Finding the intersection of a line and a circle.
5. Finding the intersection of two circles.

If we have lengths a and b in the diagram opposite, then we can construct lengths corresponding to:

$a + b$, $a - b$, $a \times b$, and $a \div b$.

The diagram shows how $a + b$ is constructed, by putting lengths end to end. We can also construct a length of $\sqrt{a}$. And that is it. It can be shown that any constructible length must be built-up using the four basic operations of arithmetic and taking square roots.

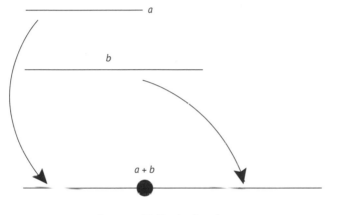

From a *and* b, *construct* a + b.

The way was clear to show that the Greek problems were insoluble. If a problem involves a length which cannot be built up from 1 by +, −, x, ÷, and √, then it cannot be constructed and the problem cannot be solved.

Doubling the Cube and Trisecting the Angle Revisited

Pierre Wantzel (1814–1848)

Using straight edge and compasses only, it is impossible to construct a cube with exactly twice the volume of a given cube, or to trisect certain angles.

Doubling the cube and trisecting the angle are two of the three famous Greek problems: both are impossible.

Any length that can be constructed using the two basic instruments must be built up using only the operations of $+$, $-$, $\times$, $\div$, and $\sqrt{}$.

The problem of doubling the cube comes down to constructing a length of $\sqrt[3]{2}$. This is a solution of $x^3 - 2 = 0$, which is a cubic equation and cannot be broken down any further. Its solutions cannot be written in terms of $+$, $-$, $\times$, $\div$ and $\sqrt{}$ (this is a consequence of Galois theory).

This length cannot be constructed using the basic instruments: it is impossible to "double the cube."

Some angles can be trisected, For example, an angle of 90° can be trisected into three angles of 30°. There is no method that works for all angles – an angle of 30°, for instance, cannot be cut into three angles of 10° each.

Suppose we *could* construct an angle of 10°. Then we could construct a length of sin 10°. This number is a solution of a cubic equation that cannot be broken down into simpler equations. The solutions cannot be expressed in terms of $+$, $-$, $\times$, $\div$, and $\sqrt{}$, hence the length sin 10° cannot be constructed.

The conclusion is that an angle of 10° cannot be constructed and the angle of 30° cannot be trisected.

The third problem, squaring the circle, resisted resolution for a few more decades.

Quaternions

William Hamilton (1805–1865)

These are a set of numbers that extends to four dimensions.

Quaternions can model positions in four dimensions. Unfortunately, in this system it is no longer always true that $a \times b = b \times a$.

Complex numbers can be used to represent positions in two-dimensional space. If a number is $a + ib$, then a represents its distance along the x axis, and b the distance up the y axis. A great deal of mechanics in two dimensions uses complex numbers.

Our world is three dimensional and it would be very handy if there could be "super-complex" numbers that work in three dimensions. Hamilton's quaternions were designed to do just that. A three-dimensional position involves three axes instead of two: the x, y, and z axes.

The ordinary bit and the complex bit of a number represent the distances along the x and y axes. Invent the "super-complex" number j, then a cj term represents the distance along the z axis.

As a bonus, they extended to four dimensions with another "super-complex" number k. The new numbers, j and k, share with i the property of being square roots of –1. A quaternion is, therefore, of the form:

$$a + bi + cj + dk,$$

where a, b, c, and d are ordinary numbers, and i, j, and k are all roots of –1. The following are the defining equations of quaternions:

$$i^2 = j^2 = k^2 = ijk = -1.$$

For ordinary numbers, and indeed for complex numbers, it does not matter in which order numbers are multiplied. For all numbers x and y, $x \times y = y \times x$.

For the algebra of quaternions to work this rule has to be abandoned. In fact:

ij = k, but ji = –k.

For this and other reasons, quaternions were not very useful in representing points in three-dimensional space. However, they have found applications in other fields, such as describing rotations in computer graphics.

Hamilton discovered the defining equations described earlier while crossing a bridge in Dublin, Ireland, and was so excited that he scratched them into the stone. Nowadays, a plaque in his honor covers up what would now be considered vandalism.

See: *Complex Numbers,* pages 101–102

Transcendental Numbers

Joseph Liouville (1809-1882)

*A number is transcendental if it is not a solution
of an algebraic equation.*

Transcendental numbers are those numbers that are not the solutions of algebraic equations. In particular, "e" is one of these.

Some numbers, such as √2, are irrational in that they cannot be written as a fraction. But they may still be found as the solution of an equation. For example √2 is a solution of the equation:

$$x^2 - 2 = 0.$$

This number, √2, is called algebraic. In general, an algebraic number is the solution of an equation in powers of x, in other words, a polynomial equation.

Numbers that are not algebraic are called transcendental.

In 1844, the mathematician Joseph Liouville constructed the first transcendental numbers. These numbers were defined solely for the purpose of being transcendental and do not have other important properties.

Very many other numbers are transcendental: indeed in a certain sense, the majority of numbers are transcendental.

The next task was to show that other numbers, which are significant in mathematics, are transcendental. In 1873 this was done for "e" by Charles Hermite.

See: *Irrational Numbers*, page 19; *e*, page 103

Kirkman's Schoolgirl Problem

Thomas Kirkman (1806–1895)

*This problem was concerned with arranging
objects in groups of three.*

This problem is a particular example
of a general dilemma to do with
arranging finite sets.

In recent years *Sudoku* puzzles have
become very popular. The compiler of a
Sudoku puzzle needs to put the digits 1
to 9 into a 9 by 9 grid, so that no digit
occurs twice in any row, column, or any
3 by 3 square. This is a problem in
combinatorics, the mathematical topic
of how to arrange finite sets so that
they obey a certain property.

An early example of combinatorics
appeared in a magazine in 1850. It is
known as Kirkman's schoolgirl problem.
It goes as follows:

*Fifteen young ladies in a school
walk out three abreast for seven
days in succession: it is required to
arrange them daily so that no two
shall walk twice abreast.*

Thus, the 15 girls must be arranged
in triples (groups of three), 7 times over.
No two girls can appear in the same
triple more than once. There are
essentially 7 ways of arranging it:
one is shown in the table opposite in
which the girls are denoted as A through
to O.

This recreational puzzle was a
popularization of a general problem
in the arrangement of finite sets.

Given n numbers (n an odd multiple
of 3), arrange them in triples in m lists,
so that no pair of numbers occurs more
than once in the same triple. This is
possible if:

m is $^1\!/_2$ $(n-1)$ or less.

Sun	Mon	Tue	Wed	Thu	Fri	Sat
AFK	ABE	BCF	EFI	CEK	EGM	KMD
BGL	CDG	DEH	GHK	DFL	FHN	LNE
CHM	HIL	IJM	LMA	GIO	IKB	OAH
DIN	JKN	KLO	NOC	HJA	JLC	ACI
EJO	MOF	NAG	BDJ	MNB	OAD	FGJ

The above table shows one way in which to arrange the school girls.

Notice that:

$\frac{1}{2}(15 - 1)$ is 7

—that is, the number of days in a week.

This subject is also important in geometries in which there are finitely many points.

The Laws of Thought

George Boole (1815–1864)

In the laws of thought, logic is reduced to algebra.

Boole set up a system to codify logical argument as a form of algebra. This started a whole branch of mathematics and is also used in the design of computer circuitry.

Logic had traditionally been a branch of philosophy. Aristotle's *Organon* had codified logic in terms of syllogisms – short three-lined arguments. One is that:

> *All men are mortal.*
> *Socrates is a man.*
> *Therefore Socrates is mortal.*

Boole noticed that many of the characteristics of logic were similar to those of algebra. He sought to set up a branch of algebra to express logical argument.

The result is Boolean algebra. In modern notation, $\wedge$ represents *and*, $\vee$ represents *or*, $\sim$ *not*, and $\Rightarrow$ *implies*.

There are various rules and axioms for this algebra, for example:

$$A \Rightarrow B \equiv \sim A \vee B.$$

"A implies B is equivalent to not A or B."

The reason is as follows. Suppose we have $\sim A \vee B$. Then if $\sim A$ is false, then B must be true (as either $\sim A$ or B is true). However, "not A" being false is the same as A itself being true. Hence A being true implies B is true; in other words $A \Rightarrow B$.

A use of Boolean algebra, which he cannot possibly have anticipated, is in the design of computer and relay circuits.

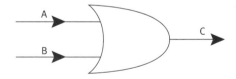

This diagram shows an AND gate. It will transmit a signal at C only when signals arrive at both A and B. So, $C = A \wedge B$.

With AND gates (and OR gates, and NOT gates) a circuit can evaluate any logical statement we want.

George Boole came from a poor background and was self taught, apart from some basic instruction from his father. He discovered his early results, in fields other than logic, while working hard as a schoolmaster to support his many dependants. A few leading mathematicians of the day recognized his worth and obtained for him a professorship at Queen's College in Cork, Ireland. Boole provides an encouraging example of intellectual rags to riches.

Twisted Shapes

August Möbius (1790–1868)
Felix Klein (1849–1925)

A Mobius strip and the Klein bottle are examples of twisted shapes, one of which (the Klein bottle) exists only in four dimensions.

A Möbius strip has just one edge and one surface. Klein bottles have no distinction between the inside and outside surfaces.

Take a rectangular strip of paper, twist it through half a turn then join the ends of the paper together. The result is a Möbius strip.

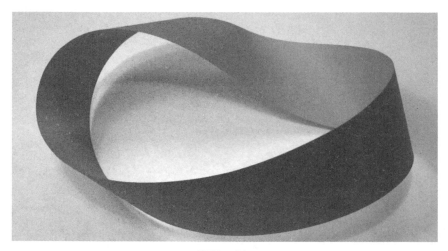

Möbius strip.

Trace along an edge. You will find that when you return to the starting point you have visited every part of the edge, on both sides of the paper. Similarly, trace along a surface. When you return to the starting point you will have covered every part of the surface, top and bottom. So the Möbius strip has the property that it has only one edge and only one surface.

Take two Möbius strips, and join them edge to edge. (Do not try this as it is only possible in four dimensions.) The result is a Klein bottle. Another way to construct the Klein bottle is to take a flexible tube, twisting one of the circular ends so that it runs anti-clockwise rather than clockwise (this is possible in four dimensions, but not in three), joining it with the other circular end.

Klein Bottle.

The Klein bottle has no inside or outside, so it cannot contain liquid. The picture above is actually misleading. It is the best we can do in three dimensions, however. The illustration shows the Klein bottle intersecting with itself, and it does not do this in four dimensions.

The Riemann Hypothesis

Bernhard Riemann (1826–1866)

This is a hypothesis about the zeros of an infinite series.

The Riemann hypothesis is one of the most famous, and perhaps the most important, unsolved problem of mathematics.

Mathematics is a highly technical subject with many of its areas incomprehensible to non-mathematicians – and even to mathematicians working in a different area. The Riemann hypothesis exemplifies this. This entry is much more technical than others in this book.

The Riemann zeta function is defined by:

$$\zeta(s) = \sum_{r=1}^{\infty} \frac{1}{r^s}$$

So $\zeta(s) = \frac{1}{1^s} + \frac{1}{2^s} + \frac{1}{3^s} + \frac{1}{4^s} + \dots$

(ζ is the Greek letter pronounced zeta.)

Some values of the function are as follows:

$\zeta(1)$ is infinite, $\zeta(2) = \frac{\pi^2}{6}$

$$\zeta(4) = \frac{\pi^4}{90} \quad \zeta(6) = \frac{\pi^6}{945}$$

These results are (comparatively) easy to prove. The function becomes important when we allow s to take complex values.

The Riemann hypothesis is:
Whenever $\zeta(s) = 0$, the real part of s, if positive, is $\frac{1}{2}$.

The Riemann hypothesis surfaces again in other areas of mathematics. Very often the solution to an unsolved problem could be found, provided that the Riemann hypothesis is true.

Goldbach's conjecture and Fermat's last theorem can be easily understood with a fairly basic knowledge of mathematics. The Riemann hypothesis is far more important than these two as its solution provides answers for many other unsolved problems.

Maxwell's Equations

James Maxwell (1831–1879)

Maxwell devised a set of equations summarizing electricity and magnetism.

Electrical and magnetic fields are propagated through space at the speed of light.

A wire carrying an electrical current creates a magnetic field (discovered by French physicist André-Marie Ampère (1775–1886) during the 1820s). If the wire is near a magnet, there will be a force on the wire. That is the principle behind the electric motor.

If a wire is moved in a magnetic field, then a current will be made to flow in the wire – discovered by British chemist and physicist Michael Faraday (1791–1867) in 1831. That is the principle behind the electric generator.

These two facts can be interpreted in terms of equations, giving the rates of change of electrical and magnetic fields. Maxwell took these equations concerning electricity and magnetism, together with others, and analyzed them mathematically. He found that the rate of change of electrical and magnetic fields is such that these will be propagated through space as waves.

So he had deduced that there are electromagnetic waves even though he hadn't actually produced them. He then found the speed of these waves using known constants. It turned out that the speed was 186,000 miles per second – the speed of light. This could not be coincidence; the two waves, of light and of electromagnetic fields, must be the same thing.

What a moment! Maxwell had shown that light consisted of electromagnetic waves. At a stroke he had reduced an important branch of science, the study of light, to a sub-branch of another.

The Countability of Fractions

Georg Cantor (1845–1918)

Cantor proved that fractions can be counted.

Every whole number is a fraction, but not every fraction is a whole number. But, in a certain sense, there are as many whole numbers as there are fractions.

The natural numbers are the counting numbers, 1, 2, 3, 4, and so on.

Include 0 and negative numbers, and we obtain the integers:

... −2, −1, 0, 1, 2, 3, ...

A rational number, or fraction, consists of one integer divided by another, such as $^5/_6$, or $^{-17}/_3$. Every integer is automatically a rational as the bottom of the fraction can be 1. The number 17 can be written as $^{17}/_1$.

An infinite set is countable if you can list its elements as a_1, a_2, a_3, and so on. Clearly the natural numbers 1, 2, 3, ... are countable. So also are the integers. You can list these as:

0, 1, −1, 2, −2, 3, −3, and so on.

Every natural number is an integer. Not every integer is a natural number. Yet the two sets, of natural numbers and integers, can be laid out alongside each other, pair by pair.

- 1 2 3 4 5 6 7 ...
- 0 1 −1 2 −2 3 −3 ...

The infinite list on the top will include all natural numbers. The infinite list on the bottom will include all integers. The two lists are paired together, so that there are as many numbers in the top list as there are in the bottom list. The bottom list, of integers, is countable.

The infinity of the integers is the same as the infinity of the natural numbers. This infinity is denoted by $\aleph_0$.

	1	2	3	4	5	6
1	$1/1$	$2/1$	$3/1$	$4/1$	$5/1$	$6/1$
2	$1/2$	$2/2$	$3/2$	$4/2$	$5/2$	$6/2$
3	$1/3$	$2/3$	$3/3$	$4/3$	$5/3$	$6/3$
4	$1/4$	$2/4$	$3/4$	$4/4$	$5/4$	$6/4$
5	$1/5$	$2/5$	$3/5$	$4/5$	$5/5$	$6/5$
6	$1/6$	$2/6$	$3/6$	$4/6$	$5/6$	$6/6$

Counting fractions.

A little more work shows that the set of fractions is countable. Arrange them in a table, then count across the diagonals. The counting of positive fractions is shown in the table above. Therefore, the listing will go as:

$$\tfrac{1}{1}, \tfrac{1}{2}, \tfrac{2}{1}, \tfrac{3}{1}, \tfrac{2}{2}, \tfrac{1}{3}, \tfrac{1}{4}, \tfrac{2}{3}, \tfrac{3}{2}, \tfrac{4}{1},$$

and so on.

There is a lot of repetition in this list but that does not matter. Every positive fraction will appear in the list sooner or later. The set of positive fractions is countable; in other words it has the same sort of infinity as the natural numbers, $\aleph_0$.

Treating infinity almost as if it were a number was highly controversial and Cantor suffered unusually vicious attacks. One fellow mathematician accused him of being a "charlatan," a "renegade," and a "corrupter of youth."

The Uncountability of the Reals

Georg Cantor (1845–1918)

Cantor proved that real numbers are uncountable.

There are different levels of infinity. The infinity of the real numbers is strictly greater than the infinity of the whole numbers.

Cantor defined an infinite set to be countable if you can list its elements. The natural numbers 1, 2, 3 ... are countable. So also are the integers and the fractions.

The real numbers can be thought of as all numbers with a decimal expansion, such as 3.58274.... This expansion may terminate or it may continue for ever. Is the set of all real numbers countable?

Suppose they are countable and can be listed. The listing might go as follows:

1st number	3.**2**94759...
2nd number	5.2**6**8370...
3rd number	8.37**1**541...
4th number	0.387**9**28...

In this array, the digits along the diagonal are shown in **bold** type. Go along this diagonal, changing each of these bold digits. We might change the **2** to a 3, the **6** to a 5, the **1** to a 2, the **9** to an 8, and so on. We obtain a new number, 0.3528, which differs from the first number in the first decimal place, from the second number in the second decimal place, and so on. It is different from all the numbers in the listing, so it cannot appear in the listing.

This is our contradiction. Any attempt to list the real numbers will always leave one out. The real numbers are not countable.

The infinity of the real numbers is strictly greater than the infinity of the natural numbers. This new infinity is written as $2^{\aleph_0}$.

On a personal note, it was this result, encountered at school, that persuaded the author of this book to study mathematics at university.

Squaring the Circle Revisited

Ferdinand von Lindemann (1852–1939)

π is transcendental.

The ancient Greek problem of squaring the circle is impossible.

The third of the three great problems of Greek mathematics took longer than the other two to resolve. In 1837 it had been shown that doubling the cube or trisecting the angle is impossible. To square the circle, that is, if you are given a circle, to construct a square with the same area, was shown to be impossible almost 50 years later.

Using straight edge and compasses only, the only lengths that can be constructed are built up using +, –, x, ÷, and √. The length must be the solution of an algebraic equation that can be found only by these five operations. In particular, the length must be algebraic. The problem was solved when Lindemann showed, in a long and complicated proof, that π is transcendental – in other words, it is not the solution of an algebraic equation.

This problem, like the other two Greek problems, has attracted an extraordinary number of false proofs. In the 1770s, both the Paris Académie des Sciences and the Royal Society in London refused to looked at any of these proofs. Yet, proofs of the construction continued to be produced, even after it was shown to be impossible.

The construction cannot be done in ordinary Euclidean space. A final twist to the story is that it can be done in non-Euclidean space.

See: *Squaring the Circle*, pages 28–29

The Correlation Coefficient

Francis Galton (1822–1911)

*This formula measures how closely
connected two quantities are.*

When data seem to show a
connection between two
quantities, is that a sign of a genuine
connection, or could it be the result of
pure chance? The correlation coefficient
gives a rigorous answer.

In many fields of science we want to know
whether there is a connection between
two things. Consider the following:

- **Medicine**. Is there a connection
  between meat eating and bowel
  cancer?
- **Psychology**. Is there a connection
  between extraversion and
  schizophrenia?
- **Sociology**. Is there a connection
  between social class and longevity?

In all these cases, the investigator will
collect and tabulate data. When a graph

is drawn of one quantity against another,
will it show a connection? Will it show
a collection of points going upwards, as
shown in the diagram (*opposite*)?
Or going downwards, or will there
be no discernable pattern at all? Even
if a pattern is spotted, there is always
the possibility that any supposed
connection is the result of random
fluctuations.

The correlation coefficient is designed
to measure these connections. It is a
mathematical function of the data that is
positive for a positive connection (as one
quantity increases, so does the other)
and negative for a negative connection. A
test of the value of the function can show
whether the connection is significant, or
whether any apparent connection could
be just a matter of chance.

The coefficient is sometimes called the
Pearson correlation coefficient, though it

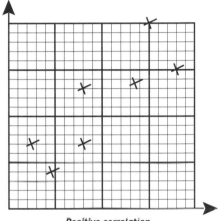

Positive correlation.

was invented by British psychologist Francis Galton (1822–1911) in the 1880s. Galton's reputation has been tarnished by his association with eugenics, a scientific movement that advocates improving humankind though selective breeding and the application of genetics – in this way superior races would survive and inferior races would die out.

The Continuum Hypothesis

Georg Cantor (1845–1918)
Kurt Gödel (1906–1978)
Paul Cohen (1934–2007)

The infinity of the natural numbers is written $\aleph_0$.
That of the real numbers is written $2^{\aleph_0}$.

The continuum hypothesis holds that there is no infinity between the infinities of natural and real numbers.

On pages 130–132 we defined the infinities: $\aleph_0$, the infinity of the natural numbers, and $2^{\aleph_0}$, the infinity of the real numbers. So, $2^{\aleph_0}$ is strictly greater than $\aleph_0$. The next infinity after $\aleph_0$ is written $\aleph_1$, then $\aleph_2$, and so on. Is $2^{\aleph_0}$ the next one up? In other words, is $2^{\aleph_0} = \aleph_1$?

Or, can you fit in another infinity between $\aleph_0$ and $2^{\aleph_0}$? That $2^{\aleph_0}$ and $\aleph_1$ are equal was hypothesized by Cantor in 1890. He tried again and again to prove it true.

It was shown to be consistent by Gödel in 1940; you cannot disprove it,

in other words. So, if you assume $2^{\aleph_0} = \aleph_1$, there will be no contradiction.

In 1963 Cohen showed that it was independent, that you cannot prove it. So if you assume $2^{\aleph_0} \neq \aleph_1$, there will be no contradiction. Indeed, $2^{\aleph_0}$ could be $\aleph_2$, or $\aleph_3$, or $\aleph_n$ for any n.

The latter years were sad for both Cantor and Gödel. Both suffered from mental illnesses with Cantor spending more time trying to prove that Francis Bacon wrote the plays of Shakespeare than on mathematics. Gödel became convinced people were trying to poison him. Towards the end of his life he produced an incomprehensible handwritten paper setting out his reasons for thinking that $2^{\aleph_0} = \aleph_2$.

Space-Filling Curves

Giuseppe Peano (1858–1932)

*There is a continuous curve that fills
a two-dimensional shape.*

A **curve has one dimension and a
square has two. Yet a curve can be
found which visits every point in the
square.**

A curve is one dimensional and is
infinitesimally thin. It seems to cover
zero area. In a highly counterintuitive
result, Peano constructed a curve which
is continuous, without any breaks in it,
which passes through every point of a
square, even though a square is two
dimensional and has non-zero area.

This curve is called space filling.
Several were invented: the following
was devised by German mathematician,
David Hilbert (1862–1943).

Divide a square into four smaller
squares and join the centers of these
smaller squares with a U-shaped curve.

Then divide each of the smaller
squares into four yet smaller squares
and join up the centers as before,
repeating the diagram four times
over. There are now four U shapes;
join them up.

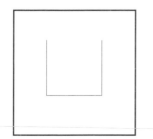

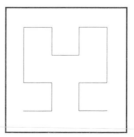

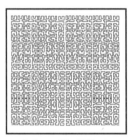

Previous page and above: *The Hilbert space-filling curve.*

And so on. The next three stages are shown in the diagrams above.

The space-filling curve is the limit of these curves. It passes through every point of the square. Though none of the original curves cross themselves, the final space-filling curve crosses itself infinitely often.

Similarly, a curve can be found which fills a solid, three-dimensional cube.

Wallpaper Patterns

Egraf Stepanovitch Fedorov (1853–1919)

A wallpaper pattern is a regular arrangement of shapes, repeated across a plane. Precisely 17 wallpaper patterns exist.

A **pattern repeats itself regularly. By classifying the operations which leave a pattern unchanged, it can be shown that there are 17 of them.**

All wallpaper patterns repeat themselves regularly. If they are moved horizontally or vertically (a move called a translation) they will remain unchanged. Some patterns are unchanged if they are rotated and some are unchanged if

they are reflected in a fixed line. These types are illustrated by the three patterns shown below.

All three patterns are built up from the same basic triangle, In the top left of the pattern. In the first diagram the triangle has been copied down and across.

In the second diagram it has been rotated through 180°, then the pair of triangles has been copied down and

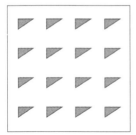

(a) Unchanged only after translation.

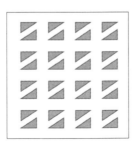

(b) Unchanged after rotation.   *(c) Unchanged after reflection.*

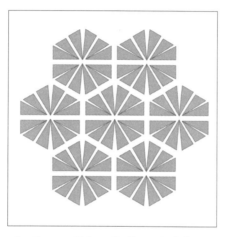

A pattern combining reflection and rotation.

across. In the third, the triangle has been reflected in a horizontal line, then the pair has been copied down and across.

If all possible combinations of translations, rotations, and reflections are considered, there are essentially 17 different types of pattern. One of the more complicated ones is shown above.

The result was discovered by Russian geologist and crystallographer E. S. Fedorov in 1891. In his work "Symmetry of Regular Systems of Figures," Fedorov proved that there were 17 symmetries in wallpaper groups. At the same time, the German mathematician, A. Schoenflies was working on the same thesis, as was English-born William Barlow.

The three-dimensional case is much more complicated: as it turns out, there are essentially 280 possible patterns.

"Doughnuts and Coffee Cups"

Henri Poincaré (1854–1912)

*Topology considered an entirely
qualitative analysis of shapes – in which doughnuts
can be considered the same as coffee cups.*

Topology studies the properties of shapes which are independent of any measurement of distance.

Traditionally, geometry was concerned with lengths, areas, volumes, and so on. By contrast, topology is involved with the qualitative, as opposed to the quantitative, properties of shapes. It deals with how many holes a solid has rather than the size of those holes.

An early example of this sort of concern was Euler's solution of the Königsberg Bridges problem. All that matters is how the bridges are connected to the land, not how long the bridges are.

The subject is often described in this way: *"Topologists think that a doughnut*

is the same as a coffee cup," In topology, two shapes are the same if one can be continuously deformed into the other. Imagine a doughnut shape made of clay. It could be squeezed into the shape of a coffee cup, without making any new holes or closing the original hole. By contrast, a doughnut is not equivalent to a pair of scissors, which has two holes.

Of course, mathematicians are not content with objects in two and three dimensions. They extend the definitions of topology to four or more dimensions. Working in four dimensions is harder than two or three, of course, but it is also harder than five, six, and so on, which is rather strange.

What the Tortoise Said to Achilles

Charles Lutwidge Dodgson (also known as Lewis Carroll; 1832–1898)

In a paradox concerning implication, the tortoise traps Archilles into an infinite regress of implication.

After the race between Achilles and the tortoise (which of course the tortoise won), they have an argument about mathematics. This starts with a simple deduction about a triangle with two sides, each measuring 12.7cm or 5in long (A). Achilles tries to convince the tortoise that the triangle is isosceles, that is that the triangle has two equal sides (B).

A   Two sides of this triangle are equal to the same number.

A⇒B   Things which are equal to the same thing are equal to each other.

B   Therefore two sides of this triangle are equal to each other.

In logical form this is as follows:

> *If A implies B is true,*
>   *and A is true,*
>   *then B is true.*

Using the $\Rightarrow$ symbol to mean "implies," then:

$$((A \Rightarrow B) \text{ and } A) \Rightarrow B \qquad (C)$$

This is called 'modus ponens' and is the basic logical rule for deduction.

The tortoise is doubtful about this proof. She still will not accept B. Therefore, Achilles asks her to accept modus ponens itself as an axiom. The argument is now:

$$(C \text{ and } (A \Rightarrow B) \text{ and } A) \Rightarrow B$$

In full:

$$((((A \Rightarrow B) \text{ and } A) \Rightarrow B) \text{ and } (A \Rightarrow B) \text{ and } A) \Rightarrow B \qquad (D)$$

The tortoise is still sceptical. Will she accept D as a further axiom, and hence will she accept this?

$$(D \text{ and } C \text{ and } (A \Rightarrow B) \text{ and } A) \Rightarrow B$$

In full:

$$(((((A \Rightarrow B) \text{ and } A) \Rightarrow B) \text{ and } (A \Rightarrow B) \text{ and } A) \Rightarrow B) \text{ and } (((A \Rightarrow B) \text{ and } A) \Rightarrow B) \text{ and } (A \Rightarrow B) \text{ and } A) \Rightarrow B \qquad (E)$$

Achilles saw, with sadness, that this could continue backwards for ever. The tortoise would never believe B was true.

This argument may sound pointless but it does raise the distinction between formal logical deduction and common sense reasoning. Where does one begin and the other end?

The mathematician Charles Lutwidge Dodgson, who invented this puzzle, is better known as Lewis Carroll, the author of *Alice in Wonderland* and *Alice through the Looking Glass*.

There is a story that the British Queen Victoria (1819–1901; reigned 1837–1901) was so delighted by the Alice books that she asked Dodgson to send her a copy of his next publication. She was surprised, and probably disappointed, to receive *An Elementary Treatise on Determinants*.

The Prime Number Theorem

Jacques Hadamard (1865–1963)
Charles de la Vallée Poussin (1866–1962)

This is a formula for the density of primes.

As we take larger numbers the proportion of primes decreases, following a logarithmic pattern.

The proof that there are infinitely many primes appears in Euclid's *Elements*. But though the sequence of primes goes on for ever, it thins out. As we take larger and larger numbers, the primes become more rare.

There are four primes up to 10 (2, 3, 5, 7). In other words, 40 per cent of numbers up to 10 are prime. There are 25 primes up to 100, so 25 per cent of numbers up to 100 are prime. Letting $\pi(n)$ be the number of primes

n	$\pi(n)$	$\pi(n)/n$
10	4	0.4
100	25	0.25
1,000	168	0.168
1,000,000	78 498	0.078
1,000,000,000	50 million	0.05
1,000,000,000,000	38 billion	0.038

up to n, the ratio of $\pi(n)$ to n decreases as n increases. The table opposite gives the proportion of primes for certain values.

We see that the ratio gets smaller. The prime number theorem is that this ratio $\pi(n)/n$ follows a logarithmic pattern. To be precise, as n gets larger, $\pi(n) \ln n /n$ approaches one. (Here $\ln n$ is the logarithm to the base e.)

To check: the value of $\pi(n) \ln n /n$ is 1.08 for $n = 1$ million, 1.04 for $n = 1$ billion, and so on. The ratio is therefore tending to 1.

The prime number theorem was conjectured in the late 18th century but was not finally proved until the late 19th century. French mathematician Jacques Hadamard and Belgian mathematician Charles de la Vallée Poussin proved it quite independently of each other – and in the same year. As a further coincidence, both men lived until well into their 90s.

The proof of the theorem is both difficult and advanced, and brings in many notions that seem unconnected with primes and logarithms. An elementary proof was found by the Hungarian-born mathematician Paul Erdös (1913–1996) in 1949. Note that "elementary" does not mean easy. The proof is still very difficult, but it does not involve advanced mathematics.

Hilbert's Problems

David Hilbert (1862–1943)

*These are a collection of problems that helped decide
the course of 20th-century mathematics.*

David Hilbert had an overview of the whole of mathematics, and in a famous speech of 1900, he identified the areas that should be addressed.

People in the 19th century witnessed unprecedented successes in mathematics – more, it is claimed, than during all the centuries before that time. For example, consider the ancient problems of doubling the cube, trisecting the angle, squaring the circle, the fifth postulate, and solving polynomial equations: all of these problems were solved in the years between 1800 and 1899.

At the end of the 19th century an International Congress of Mathematicians took place in Paris. This provided experts with an opportunity to analyze the achievements of that century and also gave them the chance to discuss what could be achieved in the future.

David Hilbert, one of the leading mathematicians of the day, gave an address listing unsolved problems, which became extremely influential in directing the way in which mathematics developed over the next century: To solve a Hilbert problem became the highest ambition of any mathematician.

Problem 10 aimed to find a method to solve any Diophantine equation (an equation which involves only the four basic operations of arithmetic). Matiyasevich showed that no such method exists.

Problem 8 looked at the Riemann hypothesis and Goldbach's conjecture. Both remain unsolved. Hilbert himself said that if he were to fall asleep for 1,000 years, his first words on awakening would be *"Has the Riemann hypothesis been solved?"*

Quantum Mechanics

Max Planck (1858–1947)

*Quantum mechanics are the physics
of subatomic particles.*

**Electrons and other particles obey
laws very different from those of
visible matter.**

Does light consist of particles or of
waves? Results in the 19th century, both
experimental and theoretical, seemed to
show conclusively that it is a wave. The
modern answer, found in the early part
of the 20th century, is that light is a bit
of both.

Planck showed that the energy of
light is not divisible indefinitely. There
is a smallest possible amount of light,
called a quantum. The energy E of a
quantum is given by the equation
$E = hv$, where v is the frequency of
the wave and h is Planck's constant, a
very small number indeed. You cannot
cut a quantum up into anything smaller
and so any emission of light must
consist of a whole number of quanta.

This was the beginning of quantum
mechanics.

When it was found that atoms are
divisible into smaller particles, they
were first treated by classical, (or
Newtonian) mechanics. The electrons
whizzing round the nucleus were
analyzed in the same way as planets
whirling round the Sun.

It was found that subatomic particles
do not obey Newton's laws. Two
differences are as follows:

1. A planet going round the Sun can, in
   theory, have any orbit of any given
   radius, and be any distance from the
   Sun. By contrast, electrons must be in
   certain fixed positions relative to the
   nucleus. They can jump between two
   orbits but they cannot occupy an orbit
   halfway between. This is because the
   energy of the electron cannot be

continuously divided – it comes in these discrete packages, quanta, which cannot be split up.

2. Classical mechanics is deterministic. If we know the exact positions and velocities of bodies at any given time we can, in theory, work out where these bodies will be at any time in the future. Quantum mechanics is non-deterministic and so there is always some uncertainty about the future behavior of a particle.

By 1900, classical mechanics had remained unchallenged for two centuries. Then relativity produced different laws for very large things, and quantum mechanics for very small things. Newton's laws still hold for middle-sized things.

The Central Limit Theorem

Alexsandr Lyapusov (1857–1918)

*The mean of a large-enough sample
is approximately normal.*

For almost any distribution, if enough data is taken, the normal distribution can be used to calculate probabilities.

The normal distribution (which is known in Germany as Gaussian distribution, and in France as the Laplacian distribution after mathematicians Carl Friedrich Gauss and Pierre Simon de Laplace) provides a good model for many sets of data that cluster round a single central value.

Laplace's work first gave an insight into the Central Limit Theorem, but it was the Russian mathematician, P. L. Chebyshev, and his students, Andrei A. Marleov and Alexsandr Lyapusov, who developed it properly.

The name "normal" is rather derogatory, implying that other distributions are abnormal, or even subnormal, certainly inferior in one way or another. To some extent this is correct, in that even a non-normal distribution will become normal if enough values are taken from it.

This was how the normal distribution was found in the first place. With a binomial distribution with large enough n, the probabilities can be approximated by finding the area under the corresponding normal curve.

In general, suppose we have almost *any* distribution. Take a sample from it and find the mean (average) of that sample. If n is large, the high values in the sample will cancel out with the low values. The central limit theorem says that as the size of the sample increases, so the curve of the mean of the sample will get closer to the normal curve. This is a powerful result that reduces the work of statisticians enormously. They no longer have to consider each case separately. Instead they can reduce almost every case to the single instance of the normal distribution.

Russell's Paradox

Bertrand Russell (1872–1970)

*English intellectual Bertrand Russell's paradox
threatened the foundations of mathematics.*

This paradox, and others, rely on objects being able to refer to themselves.

"*All Cretans are liars*", said Epimenides the Cretan, and then died of laughter at the paradox he had created. If this statement is true, then it is false because it was uttered by a lying Cretan. If it is false then it is true because it was said by someone who wasn't a liar.

A similar paradox arose in 1901 when Russell was working on *Principles in Mathematics* (1903). A set is any clearly defined collection of things. A set can contain numbers, or solid objects, and it can also contain other sets. For example, the set of all soccer teams in the United States is a set consisting of sets of soccer players.

If a set can belong to another set, then there is nothing to stop a set belonging to itself, being a member of itself, or being of a set containing itself. Some sets contain themselves, some don't.

Russell defined R to be the set of all sets that *don't* contain themselves. Is R a member of R or not?

If R is a member of R, then it doesn't contain itself, and hence isn't a member of R. If R is not a member of R, then it doesn't contain itself and hence is a member of R. That is a contradiction!

This paradox was sent by Russell to German mathematician Gottlob Frege (1848–1925), the most important logician of the day, just as he was about to publish the second volume of *The Laws of Arithmetic*. Russell's paradox destroyed it. Frege had to add an appendix to his book discussing the paradox.

A less mathematical version is as follows. Some adjectives about words apply to themselves. "English" is an English word while "polysyllabic" is a polysyllabic word. Some adjectives do not apply to themselves. "French" is not a French word and "long" is not a long word. We call an adjective Russellian if it does not apply to itself, so "French" and "long" are Russellian. Is Russellian a Russellian word or not? Either choice leads to a contradiction.

The paradox was resolved by restricting the definition of sets and language so that it was no longer possible for a set to contain itself, or for a statement to refer to itself.

See: *The Laws of Thought*, pages 124–125;
Zermelo–Fraenkel Set Theory, pages 163–164

Mathematics as Part of Logic

Gottlob Frege (1848–1925)
Bertrand Russell (1872–1970)
Alfred Whitehead (1861–1947)

*Any result of mathematics can be
obtained through logic.*

**The endeavors of Frege, Russell, and
Whitehead were to show that
mathematical proof can be reduced to
logical proof. They failed.**

George Boole had brought logic into the
family of mathematics. At the beginning
of the 20th century there were attempts
to show the reverse: that all of
mathematics could be reduced to logic.

So any branch of mathematics could
start with a set of axioms, employ
certain fixed rules of deduction, and
thereby deduce any true statement of
that branch of mathematics.

Frege set up such a system which he
hoped would be sufficient to prove any
mathematical result. His system was
destroyed by Russell's Paradox.

After demolishing Frege's system,
Bertrand Russell went on to construct
one of his own in collaboration with
Whitehead. The result was published in
a book called, rather cheekily, *Principia
Mathematica*, the same title as Sir Isaac
Newton's great work. They aimed to
show, as had Frege, that any theorem of
mathematics could be proved by clearly
stated rules of logic, starting with a set
of clearly stated axioms.

This theory, known as logicism, was
shattered by Gödel's theorem, which
found a true statement of arithmetic that
cannot be deduced logically from the
axioms of arithmetic. By then Russell
had lost interest in mathematical logic,
dismayed by the amount of repetitious
work it involved. He devoted the rest of

his long life to philosophical and political concerns. He was a pacifist during the First World War (1914–1918), for which he was imprisoned, and a key figure in the campaign for nuclear disarmament in the 1960s.

See: *The Laws of Thought*, pages 124–125; *Russell's Paradox*, pages 150–151; *Gödel's Theorem*, page 169

The Snowflake Curve

Helge von Koch (1870–1924)

The snowflake curve is jagged everywhere.

The snowflake curve is a very decorative shape with many surprising properties.

Start with an equilateral triangle. Remove the middle third of each side, and replace by two sides of a smaller equilateral triangle.

You now have 12 sides. Remove the middle third of each of these sides and replace by two sides of an even smaller equilateral triangle. Continue in this way. The first four stages of this process are shown on the page opposite:

If the process is continued to infinity, the result is the snowflake curve. It is continuous without any breaks in it, but nowhere is it smooth. At every point there is a jagged projection. On any stretch of the curve you could strike a match! The curve encloses a finite area but it is infinite in length.

The curve is an early example of a fractal. There are many projections of different sizes on the curve. Enlarge one of the smaller projections and it will look exactly the same as one of the larger projections.

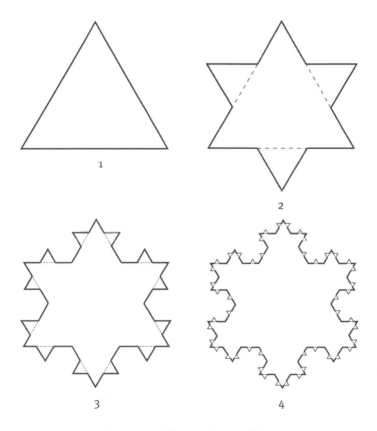

Four stages of drawing the snowflake.

Axiom of Choice

Ernst Zermelo (1871–1953)

Given a collection of non-empty sets, there is a function choosing precisely one element from each set.

The axiom of choice is used in many parts of mathematics. It is controversial because it says that infinitely many choices can be made, without giving any rule for making the choices.

A set is any collection of objects. If the set is not empty then one can pick an object from it. With two non-empty sets one can pick one object from each of them, and so this continues.

With infinitely many non-empty sets can we pick one element from each set, when there is no rule telling one how to make the choice? Bertrand Russell gave the following example: suppose the collection of sets consists of infinitely many pairs of shoes. It is easy to pick one shoe from each pair. Just choose the left shoe in each case. However, with infinitely many pairs of *socks*, there is no simple way to make the choice, as one cannot distinguish between the two socks of a pair.

The axiom of choice says that it is possible to choose one element from each set. In this case, it asserts that it is possible to choose one sock from each pair, even though there is no rule telling us which one to pick.

This axiom is important in many areas of mathematics and has many equivalents. The best known of these has the improbable name of *Zorn's Lemma* (an avant-garde film of 1970 took this as its title for no apparent reason).

The axiom does have unfortunate consequences though. Assuming the axiom, a very complicated proof of 1924 shows the following:

A solid sphere can be decomposed into six pieces, which can then be reassembled to make two spheres identical to the original.

This is the Banach–Tarski paradox (published in 1924 by Stefan Banach and Alfred Tarski). If it were physically as well as mathematically possible, one could cut a solid gold sphere into two solid gold spheres, then into four spheres, and so on, becoming indefinitely rich.

The axiom of choice can neither be proved nor disproved from the other axioms of set theory. It was shown to be consistent (unable to be disproved) by Kurt Gödel in 1940 and independent (cannot be proved) by Paul Cohen in 1963.

The Jordan Curve Theorem

Oswald Veblen (1880–1960)

*The Jordan curve theorem states that a continuous
closed curve cuts the plane into two distinct parts.*

If a curve does not cross itself and
forms a closed loop, then there are
two regions, one outside the curve and
one inside.

This theorem seems like a statement
of the blindingly obvious. If a curve
proceeds continuously without any
breaks in it and returns to its starting
point without crossing itself, then there
will be a region outside the curve and
a region inside. The two regions are
separate, one is finite and the other
is infinite.

But there are so many "pathological"
curves, such as the snowflake curve or
space-filling curves that a complete
proof is very tricky. The space-filling
curve seems impossible, just as the
Jordan curve theorem seems obvious.

After many false proofs, all of which
depended in one way or another on some
unwarranted assumption about the curve,
the theorem was finally settled in 1905.

See: *The Snowflake Curve,* pages 154–155; *Space-Filling Curves*, pages 137–138

Special Relativity

Albert Einstein (1879–1955)

Motion is only relative. There is no fixed,
stationary frame of reference.

The speed of light is always the same, regardless of the speed at which you are moving.

We receive light from all over the universe. But how does it travel? Which medium carries light waves? In the 19th century it was postulated that space is filled with a strange substance called the "luminiferous ether." It was believed that light moved in this ether, which would have provided a stationary frame of reference for all motion, including that for light waves.

Experiments showed that the laws of physics are the same, regardless of where you are and how fast you are moving. In particular, the speed of light is the same for everyone. So there is no stationary frame that is always at rest. There is no absolute motion in terms of a stationary body; there is only relative motion of one body *relative* to another.

The fact that the speed of light is always the same for everyone is hard to grasp. Light leaves the Sun at 300, 000 km/sec or 186,000 miles/sec. This speed is written c. It reaches the Earth, also at c. If a rocket is travelling at an enormous speed away from the Sun or towards it, light will still reach it at c.

The consequences of this counter-intuitive fact are very strange. A set of equations, called the Lorentz equations, predict some of the results. Suppose you are traveling towards me at a great speed. My measurement of your length, time, and mass will differ from yours. From my point of view:

- You will seem shortened in length.
- Your time will seem to be slower: in other words, your clock will appear to be running slow.

- You will seem to have a greater mass.

Of course, since motion is relative, you will think I am travelling towards you and so you will observe exactly the same things about me.

Note the last result. Your mass seems to approach infinity as your speed approaches the speed of light. Hence it is impossible to travel at the speed of light.

The increase in mass m corresponds to the energy e needed to achieve the speed, following the famous equation: $e=mc^2$.

Intuitionism

Liutzen Brouwer (1881–1966)

Intuitionism is a philosophy of mathematics that limits mathematical concepts to the human mind.

Where do mathematical objects exist? If they only exist in the human mind, then that restricts the power of mathematics.

Do the number one, the equilateral triangle, the formula $y = x^2$, and so on have an existence outside the human mind? At one extreme there is Platonism which holds that these abstract mathematical concepts do exist in some abstract universe, perhaps in the mind of God. At the other extreme there is intuitionism, which states that mathematical thoughts cannot be divorced from the human mind that thinks them.

So, for the intuitionist, the mathematician is an inventor rather than a discoverer. He or she cannot claim any external reality for any mathematical object.

This has immediate consequences in mathematical proof. Suppose one wants to prove that there is a mathematical object with a given property. For a Platonist, it is enough to show that the assumption that there is no such object leads to a contradiction. That is not enough for an intuitionist, however. One must actually produce the object with the given property.

As an example, consider this, which is relevant to public key codes. There is a certain mathematical test for whether a given number k is prime or not. If the test is negative, (in other words, the number is not prime) then the test reveals that fact and nothing more. It does not produce the numbers n and m such that $n \times m = k$, it merely asserts that these numbers, n and m, exist. A Platonist says that these factors, n and m, do exist somewhere in an abstract universe. An intuitionist believes that these factors do not exist until we can state precisely what they are.

It is much easier to do mathematics from a Platonic point of view. As a philosopher, Plato is associated with pure idealism unsullied by worldly considerations. Mathematicians, though, pull on their muddy Platonic boots during work days, while reserving their intuitionistic ideals for Sunday best.

See: *Plato and Platonism,* page 32

Zermelo–Fraenkel Set Theory

Ernst Zermelo (1871–1953)
Abraham Fraenkel (1891–1965)

These are sets defined by clearly stated axioms.

Set theory had been assailed by paradoxes. Zermelo and Fraenkel put it on a secure axiomatic foundation, clarifying what is a set and what is not.

A set is any collection of objects, abstract or concrete. Set theory was founded by Cantor, and from its beginning it had been controversial among mathematicians. Russell's Paradox and other paradoxes seemed to destroy the whole subject. The problem was that the set theorists were too generous in allowing what could be a set: they stated things such as *"The set of all sets"* or *"The set of sets that don't contain themselves,"* which gave rise to paradoxes.

Zermelo and Fraenkel took a more cautious approach. They defined sets from the bottom upwards. Starting with one simple basic set, more complicated sets can be built up from it.

What is this basic set? What object can one start with? The basic set of Zermelo–Fraenkel set theory is – nothing. There is one basic set, the set that has no members at all. This set is called the empty set and is written Ø. The axioms allow one to build more complicated sets from basic sets. We can write a set using curly brackets – the set {a, b} contains elements a and b and no others.

Starting with Ø, there is a set {Ø} which has 1 element, Ø itself. Then there is a set {Ø, {Ø}} which has 2 elements, Ø and {Ø}. We can define a set {Ø, {Ø}, {Ø, {Ø}}} with 3 elements, and one with 4 elements, or with any finite number of elements. We have a chain of sets, Ø, {Ø}, {Ø, {Ø}}, and so on. They have 0 elements, 1, element, 2 elements, and so on. So this process enables one to have a modeling of the integers 0, 1, 2, 3, ... within set theory.

There is also an axiom of infinity which permits a set containing all the elements of such a chain of sets. This set is {∅, {∅}, {∅, {∅}}, ...}. The set can be thought of as the set of integers 0, 1, 2, 3, ...

This is an important theory since the whole of mathematics can be modeled in terms of Zermelo–Fraenkel set theory, even though its contents only consist of sets built out of nothing.

See: *Russell's Paradox*, pages 150–151

The Hairy-Ball Theorem

Liutzen Brouwer (1881–1996)

If every point on a sphere is assigned a direction along the sphere, there will be at least one point with no direction.

Suppose a sphere is covered with **hair. However the hair is combed, there will always be a tuft. You cannot comb a hairy ball.**

Put more mathematically, suppose every point on a sphere is given a direction along the sphere (the direction along which the hair at that point is combed). Then there will always be at least one point where the direction is not defined, (and so there will be a parting or hair sticking out in a tuft). The proof is difficult.

The theorem has a surprisingly practical application. Consider the Earth. At each point on the Earth the wind is blowing in a certain direction. From the hairy-ball theorem, there is a point at which the wind is zero. That point is the eye of a cyclone where the wind velocity is zero. So there must always be a cyclone somewhere on Earth.

It is probably inadvisable to teach this theorem to a class of 15-year-old boys, though.

General Relativity

Albert Einstein (1879–1955)

A revolutionary theory of gravity which states that matter affects the space and time surrounding it.

Special relativity considers objects moving at constant speed relative to each other. General relativity extended the theory to acceleration (or changing speeds).

Acceleration *feels* like gravity. An astronaut in a sealed capsule, feeling the physical sensation of heaviness, would not know whether this was from the gravitational attraction of a nearby body or because the capsule was accelerating. Part of Einstein's breakthrough was to consider gravity and acceleration as being part of the same phenomenon.

The consequences are even more astonishing. The presence of matter doesn't just alter the behavior of other pieces of matter; it alters the surrounding time and space. In an extreme case, a black hole consists of matter so dense that the space surrounding it is a vortex from which light cannot escape. Another result is that space is non-Euclidean.

Although revolutionary, special relativity was not controversial. If you grant that the speed of light is the same for all observers, which is an experimentally verified fact, then everything else follows mathematically. General relativity was controversial and needed verification to become accepted. This came when an oddity in the orbit of Mercury was explained exactly by general relativity, and when, during an eclipse, light from the stars was observed to be bending around the sun.

See: *Special Relativity,* pages 159–160; *Non Euclidean Geometry*, pages 113–114

The Hilbert Program

David Hilbert (1862–1943)

This was a bold project covering the whole of mathematics.

If successful, the Hilbert program would have formalized every branch of mathematics.

In 1900 Hilbert listed 23 problems which he thought should occupy mathematicians over the 20th century. The Hilbert program was even more ambitious; in fact, it bordered on megalomania.

Mathematics, in particular the mathematics of infinite objects, had been assailed by paradoxes. The Hilbert program would eliminate these. In every part of mathematics there should be:

1. A formalization of all the axioms and rules of deduction.
2. Consistency. All mathematical deduction systems should be consistent or, in other words, unable to prove a contradiction such as "A and not A."
3. Completeness. The deduction system should be capable of proving all the statements that are true.

Take ordinary arithmetic. Task 1 creates a set of axioms for arithmetic, and a set of rules for the deduction of theorems. Task 2 would show that it is impossible to prove $1 = 2$, for example. Task 3 ensures that the deductive system could prove every true statement of arithmetic.

Furthermore, these tasks were to be carried out using only finite methods. An argument based on the infinite expansion of a decimal, for example, would not be allowed.

Suppose the program had succeeded. For mathematicians the consequences would be catastrophic. If task 3 (completeness) succeeded, there would be a formal procedure for proving any true theorem. A computer could be set to run the procedure and it could churn out theorems without the need for any human intervention. Overnight, mathematicians would become redundant.

The program did not succeed, however. In 1931 Gödel's theorem showed that there is a true statement of arithmetic that cannot be proved from the axioms of arithmetic. So task 3, completeness, is impossible. It also followed from the theorem that it is impossible to prove the consistency of arithmetic from within arithmetic itself. So task 2 (consistency proved by finite means) is also impossible.

The Hilbert program obtained many important and lasting results. Fortunately it could never achieve its original goals in their entirety.

See: *Hilbert's Problems,* page 146; *Gödel's Theorem*, page 169.

Gödel's Theorem

Kurt Gödel (1906–1978)

*There is a theorem of arithmetic which cannot be
proved from the axioms of arithmetic.*

Arithmetic is incomplete. However
many axioms are stated, there is
always a theorem which cannot be
deduced from those axioms.

Russell and Whitehead tried to show
that mathematics is a branch of logic,
in that any theorem is a purely logical
consequence of the axioms of its system.
In 1931 Gödel spectacularly proved
them wrong by producing a theorem of
arithmetic which could not be proved from
the axioms of arithmetic. So arithmetic
(and other parts of mathematics) is more
than a branch of logic.

He did this by assigning code numbers
to all the symbols and statements of
arithmetic. He produced a statement
which says:

*The statement with code number
g cannot be proved*

The code number of this statement is
g itself. So this statement refers to itself.

Suppose this statement were false.
Then the statement with code number *g*
can be proved, and hence would be true.
But this is the statement with code
number *g*! The statement is true but
unprovable nevertheless.

Gödel demolished not just the
logicism of Russell and Whitehead but
also the prediction of Leibnitz that
ultimately any argument could be settled
by mechanical calculation.

Not many theorems can be described
as life-enhancing but this is one of them.
It shows that no one can program a
computer to churn out all possible
truths. There is always the need for
human intelligence.

On a personal note, it was this result
that persuaded the author of this
book to continue with mathematics
after university.

The Traveling Salesman

Karl Menger (1902–1985)

A traveling salesman must visit a number of cities.
What is the most efficient route?

How should a traveling salesman plan his route when he sees before him a map showing all the cities he must visit? One obvious method is to visit the nearest city first, then the city nearest to that and so on. This usually gives a good route but it is not necessarily the shortest one.

The problem is that with more and more cities, the number of routes increases very rapidly.

With 2 cities there are 2 routes.
With 3 there are 2x3 = 6 routes.
With 4 there are 2x3x4 = 24 routes, and so on.
With 10 cities there are 3,628,800 routes.
With 15 cities there are over a trillion routes.

The "Brute Force" solution to the problem is to examine each of the routes in turn and pick the shortest. Even with the fastest computer, this is impossible with a large number of cities. At the moment this is an unsolved problem – there is no known method for finding the shortest route that can be done in a reasonable period of time.

A variation of this problem is the "Traveling Politician." She is touring the whole country before an election and must visit one city from each state: the problem is more complicated as she must choose which city to visit as well as the order in which to visit them.

The problem applies also in manufacturing. Imagine a machine tool that drills a number of holes in a steel plate. In which order should the holes be drilled, to minimize the time spent traveling between the holes?

Turing Machines

Alan Turing (1912–1954)

Alan Turing described machines that define what is and what is not possible in terms of computing.

A Turing machine is a computer, stripped down to its bare essentials. It is an abstract idea of a machine rather than a solid thing of metal and plastic. It envisages an indefinitely long strip of tape marked with 0s and 1s. This tape passes under a head which reads the number and then performs two tasks, one on the number and one on the tape:

1. Changes a 1 to 0 or 0 to 1 or leaves the number as it is.
2. Advances the tape by one, retards the tape by one or leaves the tape as it is.

A Turing machine can be designed to perform addition, subtraction, or any basic operation of arithmetic. In fact it can perform any arithmetical operation, however complicated. Going one step further, any computing task can be performed by a Turing machine.

It seems that there must be a separate Turing machine for every task. But in an astonishing result, he showed that you don't need separate machines for each task. Instead, there is a "universal" Turing machine that can act as any other Turing machine. So this single universal machine can perform any computable task – this is called Turing complete.

Another important result is the halting problem. Given a Turing machine, we cannot always decide if it will come to a halt or continue calculating for ever. This is the computing equivalent of Gödel's theorem.

During the Second World War Alan Turing did invaluable work in the field of decoding. He was also an unapologetic homosexual at a time when it was illegal, and was convicted for gross indecency in 1952. He committed suicide two years later, using an apple laced with cyanide, in a macabre echo of Snow White.

Binary Numbers

Gottfried Leibnitz (1646–1716)

Claude Shannon (1916–2001)

George Stibitz (1904–1995)

Binary numbers are a system of writing numbers using only two symbols, 0 and 1.

The use of binary numbers, a simple way of writing numbers, is used in computers.

Normally we use ten symbols to write numbers, the digits 0 to 9. This is called a base 10 system, as each digit in a number is multiplied by a power of 10.

A simpler system is base two, or binary. We only have two symbols, 0 and 1. In a number each digit is multiplied by a power of two. For example the number 1101 in binary represents:

$$1 \times 2^3 + 1 \times 2^2 + 0 \times 2^1 + 1 = 13$$

There were many precursors, but the first fully worked out system of binary numbers is due to Leibnitz. He was ahead of his time, though, as binary numbers only came into their own with the invention of electronic computers, which do all their calculations in binary.

This is for two reasons. The first is that computer memory is composed of units which can be in one of two states. In early computers they were elements which could be magnetized either clockwise or anticlockwise. One state represents the number 0, the other the number 1. A chain of these elements can represent a binary number. The other reason is that electronic circuitry to perform arithmetic on binary digits is much more simple than for base 10 digits.

In 1937 Claude Shannon showed how to design circuits for binary arithmetic. In the same year George Stibitz built a machine that actually did it – the machine was called Model K, as it was built on his kitchen table.

Leibnitz was a forerunner of computing in other ways. He invented a mechanical calculator. He had a very optimistic view of the scope of logical reasoning, envisaging that in the future all thought would be reduced to calculation.

... when there are disputes we can simply say: Let us calculate ... to see who is right.

See: *Calculating Machines,* pages 72–73

The Ham Sandwich Theorem

Stefan Banach (1892–1945)

*Given three solids, there is a plane cutting
each of the three solids exactly in half.*

A ham sandwich consists of a slice of ham between two slices of bread. With a single straight cut you can ensure that the ham and the two slices of bread are each cut into equal halves.

This is a rather culinary topic which starts in two dimensions with the Pancake theorem as follows:

Given two slices of pancake there is a straight line that cuts both of them exactly in half.

There are two ways you can move the cutting knife: by moving it up and down and by turning it. You can move the knife to cut the first pancake in half, then adjust the angle to cut the second.

The Ham Sandwich theorem is the three-dimensional case. It may be a badly cut sandwich with the pieces of bread and the ham not neatly placed on top of each other.

Neither do the bread and the ham have to be flat; they can be as lumpy as we like. In all cases, there is a single plane that cuts all three exactly into half.

These results extend into higher dimensions. They are special cases of a later, more general result which is called the Stone–Tukey theorem. This states that:

Given n *shapes in* n *dimensional space, there is an* (n − 1) *dimensional plane that cuts each of the shapes exactly in half.*

It is difficult to image what a four-dimensional ham sandwich would look like, let alone how it would taste.

The Enigma Machine

Marian Rejewski (1905–1980)
Alan Turing (1912–1954)

*During the Second World War (1939–1945) a way was
found to break a seemingly unbreakable code.*

During the Second World War the German army and navy used the Enigma, a coding machine, to render their messages and correspondence indecipherable to their enemies. Mathematicians based in Poland and Bletchley Park (England's famous codebreaking headquarters) found a way to break the code.

In a code, or cipher, the letters of the original message are replaced by other letters to obtain the coded message. The rule by which the letters are replaced is called the key of the cipher.

In a monoalphabetic code each letter is always substituted by the same letter. This can easily be broken by counting the frequency of each code letter. More secure is a polyalphabetic code in which the substitution order changes. So the letter "A" might be coded as "M" in one place and as "P" in another.

The Enigma machine (as made famous in the 2001 film *Enigma*) looked like a typewriter with a set of rotors at the back. As each letter was typed in, a current would pass through the rotors and cause a coded letter to emerge. The rotors would then turn to a different position so that the order for coding changed for the next letter. The configuration of rotors would not be repeated until over 17,500 letters had been typed in. The machine contained other devices which extended the number of possible keys to 10,000,000,000,000,000.

Clearly it was impossible to test them all and, unsurprisingly, the Germans thought that the code was impregnable. The key was changed daily, and even if the enemy found the key for one day it would be changed for the next.

Despite its complication, Polish mathematicians, particularly Marian Rejewski (1905–1980), succeeded in breaking the basic Enigma code before the war. He did it by analyzing the code in terms of groups of permutations. Then, unfortunately, the machine became even more complicated with different selections of rotors being made possible.

In 1939 Poland was conquered, France was overrun the following year and the center of code-breaking was moved to Bletchley Park in southern England.

Prominent among the code-breakers was Alan Turing who used a system, pioneered by Rejewski, of harnessing several Enigma machines to test the possible settings. Eventually it became

possible to find the key for the day within an hour or so, and hence to decode all the Enigma messages for that day. At midnight, however, the key would change and work would begin again.

One success of Bletchley Park was in predicting German submarine attacks. This stemmed the enormous loss of shipping in the Battle of the Atlantic in 1943. Some historians estimate that by this and other such operations the code-breakers shortened the war by about two years.

However, despite their importance, the code-breakers were sworn to secrecy and the whole operation was kept secret until 1970. During the 25 years following the war, they had to endure in silence the taunts of all those who accused them of shirking active service.

See: *Colossus,* pages 177–178; *Galois Theory*, page 115; *Turing Machines*, page 171

Colossus

Tommy Flowers (1905–1998)

Possibly the world's first electronic computer, the Colossus was used during the Second World War (1939–1945) for decoding top-secret German messages.

The code-breakers at Bletchley Park had succeeded in decoding the secret messages sent by the Enigma machine. But they could not find a way to unlock top-secret information sent by a more complicated device called the Lorentz machine.

Enigma was used by the Germans to send coded messages of comparatively low-level importance, for example to units in the field of battle or to submarines at sea. Messages between Hitler and his generals were at a higher level of secrecy, and so were sent via the Lorentz machine. It still worked by settings of rotors on a machine but was altogether too difficult to be solved by the methods that had successfully broken the Enigma codes. The Lorentz machine had just too many possible settings.

To solve this dilemma, Tommy Flowers, a telecommunications and electronics expert, proposed the Colossus system, using his experience of the British telephone system. Colossus was highly sophisticated, using 1,500 valves (far more than had ever been used before) and operating five times faster than any other system.

The devices used to combat the Enigma code were electromechanical, in other words they still had parts which were physically moving. The Colossus used electronic valves that did not move at all. These electronic components were used to mimic the settings of the Lorentz machine. The intercepted code message was keyed onto a paper tape, which was fed into Colossus. The Colossus then generated messages corresponding to possible settings of the Lorentz machine.

There were two streams of data to examine: one corresponding to the intercepted message, the other to messages generated by Colossus. The streams were studied and compared and when they were sufficiently close, the code-breakers could home in on the actual settings. The electronic valves of Colossus could compare many different settings at a speed much greater than any previous code-breaking devices.

About 11 Colossus machines were built. After the war they were "destroyed into pieces no bigger than a man's hand," in the words of Winston Churchill, and knowledge of this pioneering work in computing only resurfaced in the 1970s. The Colossus was programmable, in that it could be adjusted for different settings of the code, but it was not a fully general purpose machine. It was not "Turing complete." However it still has a claim to be the world's first electronic computer.

See: *The Enigma Machine,* pages 175–176; *Turing Machines*, page 171

Game Theory

John von Neumann (1903–1957)
Oskar Morgenstern (1902–1977)

Game theory is a branch of mathematics to do with strategies in games.

Game theory applies not just to games, but also to economics, politics, and even biology.

The theory is mainly relevant to games in which each player withholds some information from the other player or players. It does not apply to games such as chess, in which there are no secrets.

What strategy should the player follow? Consider poker. Each player knows whether he or she has a good or a bad hand and this information is kept from the other players. A player who always bets high with a good hand, and always resigns with a bad hand, will be predictable and therefore easy to beat. To be a competent poker player, one must sometimes bet high with a bad hand (this is called bluffing) and sometimes bet timidly with a good hand. That way opponents are kept guessing.

One must not have a rigid rule for play. On the contrary, it must be variable. Game theory finds the best strategy. For instance, part of the best strategy for poker might be "With a bad hand, resign 70 per cent of the time, bluff 30 per cent of the time."

A zero-sum game is one for which the sum of all the gains and losses is zero. Poker is a zero-sum game, as what one player loses another player wins. Another zero-sum game, much simpler than poker, gives surprising results. This is called "Two fingered Morra."

Each of two players, X and Y, raises either one or two fingers. The payouts are:

If both raise one finger, X pays Y $1.
If both raise two fingers, X pays Y $3.
If one raises one finger and the other raises two fingers, Y pays X $2.

X raises

	1 finger	**2 fingers**
1 finger	−$1	$2
2 fingers	$2	−$3

Y raises

X's gain.

The game can be illustrated by a table (*see table above*) which gives the gain and loss for every eventuality. This table gives the gain for player X. A negative gain is a loss.

It seems perfectly fair. Indeed, if both raise one or two fingers with equal frequency, it will be fair. In fact, the game is heavily biased in favor of X. By raising one finger $\frac{5}{8}$ of the time, and two fingers $\frac{3}{8}$ of the time, he will, on average, win $\frac{1}{8}$ per game, whatever Y's strategy is. That represents quite a profit margin!

Von Neumann made outstanding contributions in many areas of mathematics and physics. His brilliance as an inventor was not matched by skill in teaching – he would write an equation in a small corner of the blackboard and then rub it out before the students had time to copy it down.

He worked on the Manhattan project, which was set up in 1942 to develop atomic bombs, and which contributed to the end of the Second World War (1939–1945) when atomic weapons were exploded over Hiroshima and Nagasaki in Japan. Contact with radioactive materials may have caused the cancer that led to his premature death at the age of 54.

ENIAC

John Mauchly (1907–1980)
J. Presper Eckert (1919–1995)

ENIAC was the first electronic general-purpose computer.

The Electronic Numerical Integrator and Computer (ENIAC) can lay the strongest claim to be the world's first electronic computer.

Babbage's analytic engine was mechanical rather than electronic and was never built anyway. The Colossus was electronic, and worked, but could only handle a limited range of tasks. The German Z3, built in 1941, was electromechanical rather than electronic. The first machine to fulfil all the requirements was ENIAC, which was electronic and general purpose.

Originally designed during the Second World War to perform calculations connected with artillery, it became fully operational in 1946. It had nearly 20,000 valves, compared with a mere 1,500 for the Colossus, and weighed nearly 30 tons. It could multiply two 10-digit numbers in less than 0.003 seconds, which is very slow for a modern computer, but which was 1,000 times faster than the calculating machines of the day.

Electronic valves are notoriously unreliable and, with 20,000 of them, experts thought that ENIAC would be continually breaking down but by keeping the machine switched on (keeping the valves at constant temperature) stoppages were kept to a minimum. Its record was five days of continuous operation. It was finally sent out to pasture in 1955.

ENIAC was originally designed for a military purpose for the US government. This was thought to be the role of computers at the time. Both Alan Turing and John von Neumann said that there would be no need for more than four or five computers in the whole world. There are now about a billion.

The Prisoner's Dilemma

Albert Tucker (1905–1995)

The prisoner's dilemma is the most famous example in game theory.

Two suspects are arrested. Should each one stay silent or confess to the crime? This dilemma also has implications for morality.

A zero-sum game is one in which the gains of one person are the losses of another. The most famous example used in game theory is the prisoner's dilemma, which is most definitely not a zero-sum game. It goes like this:

A bank is robbed, and two suspects are found in a stolen car. The police have evidence for the charge of stealing the car, but not enough for the charge of robbing the bank. The police interrogate the suspects separately and offer the following deal:

1. If you both confess, you will both get five years in jail for bank robbery.

2. If you both stay silent, you will both get two years in jail for car theft.

3. If you confess and your partner does not, you will walk free and your partner gets 10 years in jail.

4. If you do not confess and your partner does, he will walk free and you will get 10 years in jail.

Each prisoner reasons as follows:

If my partner stays silent, I will get two years by not confessing (option 2) and walk free by confessing (option 3).

If my partner does confess, I will get five years by confessing (option 1), and 10 years by not confessing (option 4).

The table opposite shows the results of all the eventualities, from the point of view of prisoner A.

Notice that for Prisoner A, the results in the Confess column are better than

Prisoner A

		Confess	Stay silent
	Confess	5 years jail	10 years jail
Prisoner B	Stay silent	Walk free	2 years jail

those in the Stay silent column. Whether Prisoner B confesses or stays silent, Prisoner A is better off by confessing. Similarly, Prisoner B works out that he too is better off by confessing. So both prisoners confess, and both get five years in jail.

Notice that if both prisoners had *not* confessed, they would only receive two-year sentences. So by following their own self interest, the prisoners have received a considerably greater penalty than they would have done had they collaborated. Discuss!

Electronic Calculators

*Cheap, portable calculators have revolutionized
the way we do arithmetic.*

The first electronic calculators were the size of a desk and very expensive. Now they fit in a pocket and cost just a few pounds or dollars.

Mechanical calculators have been in use since the 17th century. They were still around in the 1960s and still working on the same principles, though some were electrically powered.

The first *electronic* calculator was built by IBM in 1954. Casio made their first in 1957. These were all as big as a desk and it wasn't until 1967 that Texas Instruments produced the first "handheld" calculator. Its output was via a printed tape rather than on a screen. The first calculator that would genuinely fit in a pocket was the Busicom "Handy," made in Japan in 1971. It was also the first to use a light emitting diode (LED)

display, and to run on batteries rather than mains power.

Since then, calculators have rapidly decreased in size and price, and increased in capability. They are also much cheaper to run, now that liquid crystal displays have replaced LED. Scientific models can do a wide range of mathematical tasks, far beyond the four operations of arithmetic.

Concern has been expressed that the use of calculators in schools will prevent the learning of basic arithmetic skills. Just as seriously, over-reliance on calculators may hamper understanding of how numbers behave. Nowadays, it is common to see children using a calculator to add zero or to multiply by one. Many public examinations now forbid calculators for at least some of the assessment.

See: *Calculating Machines,* pages 72–73

Polya's Principles

George Polya (1887–1985)

These are a set of rules for solving mathematical (and other) problems.

The set of rules for solving mathematical problems are the following:

1. Understand the problem.
2. Devise a plan.
3. Carry out the plan.
4. Look back.

George Polya was a Hungarian who emigrated to the United States in 1940. He is most famous for his work in the field of mathematics education. While teaching, he saw his pupils often failed to get started on a problem because they lacked an organized structure.

In *How to Solve It* (1957), Polya wrote down a set of principles, listed at the top of this page, for solving mathematical problems.

The first principle makes sure that the student understands exactly what is required and provides a framework for setting about the task. To identify this:

What is the unknown? What are the data? What is the condition?

The second principle is the most important. Polya encourages the solver to devise a methodical plan for the problem rather than rely on random stabs at it. One recommendation is:

Look at the unknown! And try to think of a familiar problem having the same, or a similar, unknown.

The third principle, of carrying out the plan, should be straightforward if the plan has been chosen wisely. The recommendation is:

Check each step. Can you see clearly that the step is correct?

The fourth step should lead the student to greater success in the future.

Can you use the result, or the method, for some other problem?

Erdös Number

Paul Erdös (1913–1996)

A number which shows how closely connected you are to one of the most prolific mathematicians of all time.

Paul Erdös wrote an astonishing number of mathematical papers. The Erdös number of a mathematician shows the chain of collaboration up to Erdös himself.

Paul Erdös emigrated from Hungary to the US, but it is unfair to assign him to any single nation. He was utterly disdainful of possessions, and wandered from country to country with only a suitcase.

Throughout his life he wrote about 1,500 papers, making him one of the two most prolific mathematicians of all time (the other being Leonhard Euler of the 18th century). He was very generous with collaboration with other, lesser, mathematicians, and it is from this that the Erdös number is defined.

If you were Paul Erdös, your Erdös number would be 0. If you were to write a mathematical paper with someone with Erdös number n, your Erdös number is $n + 1$.

So, if you have collaborated on a paper with Erdös your number is 1. If you have collaborated with someone who collaborated with Erdös, (someone with Erdös number 1), your Erdös number is 2. And so on. The smaller your Erdös number the more closely you are connected with him.

Erdös also had the honor of a limerick being written about him:

A theorem both deep and profound,
Says that every circle is round.
In a paper by Erdös,
Written in Kurdish,
A counterexample is found.

Chaos Theory

Edward Lorenz (b.1917)

This is a physical or mathematical system that behaves unpredictably.

In a chaotic system a tiny change in one part can cause enormous changes elsewhere.

The behavior of a chaotic system is deterministic; in other words, it is controlled by fixed rules, whether physical or mathematical. But the behavior is so wildly unpredictable that it *looks* as if it were caused by chance.

One important property of a chaotic system is that a small alteration to the initial values may cause a great change in the long run. This is well-known as the Butterfly Effect.

"Does the flap of a butterfly's wings in Brazil set off a tornado in Texas?"

Weather patterns are chaotic, so the answer is, yes, it could. It is important to note, however, that the butterfly's wing-flap could equally well *prevent* a tornado in Texas.

The quotation above is the title of a talk given by Edward Lorenz, a meteorologist, working on weather forecasting. A computer had predicted certain results for the weather, and he decided to run the program again to check. To his surprise the results were different. He looked at his data, and found that the only difference was that he had entered a number as 0.506 instead of 0.506,127. The tiny change of writing a number to three instead of six decimal places had caused a great difference in the final prediction.

A purely mathematical example of a chaotic system is the logistic model. In applied math, the problem of working out the positions of three bodies circling each other in space often results in chaotic solutions.

Chaos occurs in many places in the natural world. Something as simple and familiar as a dripping faucet may be chaotic.

The Secretary Problem

F. Mosteller (1916–2006)

This concerns the problem of picking the best secretary from a pool of candidates.

The candidates for a secretarial post are interviewed one by one. How do you maximize the chance of picking the best candidate?

Suppose there is a single post for a secretary. A large number of candidates apply and each is interviewed. After the interview the candidate is either appointed or rejected and candidates cannot be recalled.

Let the number of candidates be n. The strategy to maximize the probability of choosing the best candidate is as follows:

reject the first r candidates and then select the next candidate who is superior to all the previous candidates.

The method sounds cruel – if you apply it to the problem of picking a secretary, the first person on the list is bound to be rejected!

If this strategy is taken, what should r be? If r is too small, then we may be making too hasty a choice, as we have seen only a few applicants. If r is too large, too close to n, then the last few candidates might be inferior to all the previous candidates. We would then appoint the very last candidate, who might be a dud.

It turns out that the best choice for r is n/e, where e is the Euler constant. The probability that the best applicant is chosen is $1/e$.

This problem occurs in many other guises, one of which is the "Sultan's dowry problem." The Sultan has to choose a bride from a number of princesses and his sole aim is to marry the princess who will bring the largest dowry. If there are n

princesses, he rejects the first n/e of them, and weds the first princess who brings a dowry greater than all of the ones he has rejected.

This method of decision making is common in real life. On holiday, when going out to eat in a strange town, people often walk past the first few restaurants and then pick the next restaurant that seems nicer than all the ones they've seen so far. Experimental psychologists have observed people in these sorts of situations. It is found that, in practice, their choice for r is less than n/e. Perhaps they feel impatient and want to end the uncertainty.

See: *e*, page 103

Catastrophe Theory

René Thom (1923–2002)

*Catastrophe theory deals with slight
changes that can lead to great changes.*

How can one explain sudden
catastrophic changes?

Most applied mathematics is continuous.
That is, a slight change in the causes will
result in a slight change in the effects.
That is why calculus is so important: it
gives a rule for how a small change in
one variable will make a small change
in another.

   This does not always happen though.
Sometimes a small change will lead to a
large change. That is called a
catastrophe. Here are two examples:

1. A bridge supports the weight of
   the vehicles crossing it. The
   bending of the bridge changes
   continuously with the weight. At a
   certain level, though, the weight
   is too much and the bridge
   collapses.

2. Suppose a species of animal
   has been living and evolving
   happily for millions of years.
   The environment has been
   changing continuously, and so
   the species has been adapting
   continuously to fit the
   environment. The environment
   passes a certain value and it
   is no longer possible for this
   type of animal to adjust.
   Either it becomes extinct or it
   evolves to a radically different
   species in order to survive.

   René Thom sought to find a
mathematical way to analyze such
catastrophic changes. He found seven
types of catastrophe. One, given
opposite, is the graph of $y = x^3 + ax$.
Three graphs are shown opposite, for
$a < 0$, $a = 0$ and $a > 0$.

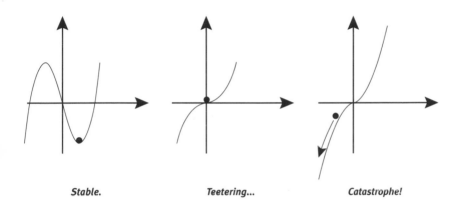

Stable. *Teetering...* *Catastrophe!*

Imagine a marble in the left-hand graph, where $a < 0$. It will rest in a stable way in the hollow of the curve. Now let a increase from negative to zero to positive. When $a = 0$, the middle graph, the marble is balancing precariously. When $a > 0$, the right-hand graph, the hollow has disappeared and the marble tumbles down the curve. The result is catastrophe!

Life

John Conway (1937–)

Life is a 'game' that has intriguing mathematical properties.

The game of life shows how populations grow indefinitely or shrink to extinction.

John Conway is one of many important mathematicians who have contributed vital results to many areas. He is best known for the game of Life which models growth and decay in populations.

Imagine an infinite square grid. Each square may or may not have a person living on it. Note that each square has eight neighboring squares. The rules for birth and death from one generation to the next are:

1. With zero or one neighbors, a person dies of loneliness.
2. With more than three neighbors, a person dies of overcrowding.
3. With two or three neighbors, a person survives.

If an empty square has three non-empty neighbors, then a person is born there (notwithstanding the fact that this process normally takes only two).

The diagram opposite shows an initial population and the next generation. Notice that the topmost and left most people have died of loneliness, and that two new people have been born on the right.

Starting with an initial population, what happens? Does it die out, grow indefinitely, or settle down to a fixed value? All of these variations can occur. Inventive phrases are used to describe the various possibilities.

For example, there are "gliders" which move across the grid, a bit like nomadic groups searching for fresh pastures. There is a "glider gun," which is a configuration that regularly sends out gliders – hence it is possible for a finite

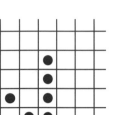

Initial population.

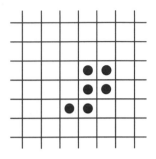

The next generation.

initial population to grow indefinitely. There are "puffers," which move across the grid leaving clumps behind them. This is another example of indefinite growth. A "Garden of Eden" configuration cannot occur as the result of any previous configuration.

The game of Life is best studied using a computer simulation. One important question to ask of any species is: will it become extinct? In terms of Life, is there a way to decide whether the population will increase indefinitely, shrink to nothing, or settle down to some fixed value? The answer is no. There is no general way to decide what will ultimately happen to a population.

In other words, Life cannot be decided.

Matiyasevich's Theorem

Yuri Matiyasevich (1947–)

Every recursively enumerable set is Diophantine.

If there is a rule for listing the elements of a set, that list can be found as the solutions of a formula involving only the addition, subtraction and multiplication of integers.

A recursively enumerable set is a set of integers that can be listed according to a fixed rule. The listing does not have to be in increasing order. Examples of recursively enumerable sets are:

1. The set of squares, $\{1, 4, 9, ...\}$.
2. The set of perfect numbers.

For the second example, just look at each number in turn and decide whether or not it is perfect. If it is, write it in the list. If it is not, leave it out and move on to the next number.

A Diophantine set is something that seems much simpler. A set is Diophantine if it can be listed as the values of a formula involving only $+$, $-$, and x. Take the examples above.

1. The set of squares is easily listed as the values of the formula x^2. So $\{1, 4, 9, ...\}$ is a Diophantine set.
2. The second example is much less obvious. Is there a formula which generates all the perfect numbers?

What Matiyasevich showed was that Diophantine sets are not simpler than recursively enumerable sets, that if there is a rule for finding the elements of a set, that search can be carried out by an algebraic formula. However complicated the rules are for listing the terms, the rules can be expressed just in terms of $+$, $-$, and x.

Many topics in this book have involved prime numbers. In the 17th century two formulae were suggested for listing primes: $2^n - 1$ and $2^{2^n} + 1$. The primes are a recursively enumerable set. Just look at each number in turn and if it is prime, write it down in the list. Matiyasevich realized that there must be a Diophantine formula for listing the primes, but he thought it would be too complicated to write down.

In 1976 the American J. P. Jones found a formula to list the primes. This used 26 variables, which was very convenient, otherwise he would have had to extend the alphabet. For the sake of completeness, the formula is written out below. As a, b, c, ..., z run through all possible numbers, the positive values of this formula run through all the prime numbers.

So here, at last, is a formula to generate prime numbers.

$$(k + 2)\{1 - [wz + h + j - q]^2 - [(gk + 2g + k + 1)((h + j) + h - z]^2 - [2n + p + q + z - e]^2$$
$$- [16(k + 1)^3(k + 2)(n + 1)^2 + 1 - f^2]^2 - [e^3(e + 2)(a + 1)^2 + 1 - o^2]^2$$
$$- [(a^2 - 1)y^2 + 1 - x^2]^2 - [16r^2y^4(a^2 - 1) + 1 - u^2]^2$$
$$- [((a \shortmid u^2(u^2 \quad a))^2 \quad 1)(n + 4dy)^2 + 1 - (x + cu)^2]^2 - [n + l + v - y]^2$$
$$- [(a^2 - 1)l^2 + 1 - m^2]^2 - [ai + k + 1 - l - i]^2 - [p + l(a - n - 1) + b(2an + 2a - n^2 - 2n -$$
$$2) - m]^2$$
$$- [q + y(a - p - 1) + s(2ap + 2a - p^2 - 2p \quad 2) \quad x]^2 \quad [z + pl(a \quad p) + t(2ap - p^2 - 1) -$$
$$pm]^2\}.$$

The formula for prime numbers.

See: *A Formula for Primes Numbers,* page 76

P = NP?

Stephen Cook (b.1939)

*This is an unsolved conjecture about
the length of computation.*

**If P=NP were shown to be true, the
secrecy of codes would be under threat.**

Computers can calculate extraordinarily
quickly, but not infinitely quickly. There
are some dilemmas, such as traveling
salesman problems, which would take
an unfeasibly long time to work out,
even on the fastest computer.

Imagine a problem in which there are
n inputs. For the traveling salesman, n is
the number of towns he must visit. As n
increases, how does the length of time
to solve the problem increase? For a
tractable problem, the length does not
increase too fast. Tractable functions
belong to class P. (P stands for
polynomial: the formal definition of
tractable is that the time increases as a
polynomial function of n.) Tractable
problems are solvable in a reasonable
time. Examples of such P problems are:

1. Multiplying together two n-digit
   numbers.
2. Given a list of n names, sorting
   them into alphabetical order.

The traveling salesman who must visit
n towns has $n!$ routes to choose from.
This function grows much faster than
any polynomial. So, to date, it is not a
tractable problem: it is not P.

Another example of a problem which is
not tractable (to date) is that of factorizing
an n-digit number. But consider this: If I
make a lucky guess for p, then to check
whether or not p is a factor of the number
takes only polynomial time. Such
problems, which can be reduced to
tractable by a lucky guess, are called NP
(for Near Polynomial).

The conjecture is that P = NP. If a
problem is tractable with a lucky guess
(NP), then it is tractable without such a
guess (P). Put another way, if the problem

can be verified quickly (by a lucky guess), then it can be solved quickly (the lucky guess can be found quickly).

If the conjecture were proved true, that would have enormous repercussions. Some modern codes are based on operations like factorizing huge numbers, and at present this is an NP problem. If it were shown to be P, then the code could be broken very easily.

The title of this topic, P = NP?, is set in the brickwork of the computing faculty at Princeton University. The brickwork can quickly be changed to P = NP! if the conjecture is proved true, and P < NP if it is proved false.

See: *The Traveling Salesman,* page 170; *Public Key Codes*, pages 198–199

Public Key Codes

Public key codes are those in which the encoding is easy while decoding is difficult.

In these codes the method of sending a message is made public. To decode the message requires extra, private knowledge.

In traditional codes the procedures for encoding and decoding are the same. In particular, both the sender of a message and its recipient must know the key of the code. In the case of the Enigma machine, the key was the setting of the machine's rotors and plugs, and this setting had to be sent to the recipient. There was always the danger that the enemy would intercept the key.

In a public key code the procedures are different. The key for encoding is made public and anyone can use it to send a message. The key for decoding requires extra knowledge. So, if you have a public key code anyone can send you a secret message, but only you can decode it.

These codes were invented in the British government's communications center in 1973 (and the discovery was kept private). They were reinvented in 1976 in the US (and the discovery was made public).

The best-known is the RSA (Rivest, Shamir, and Adleman) cipher. It works roughly as follows. Pick two large prime numbers p and q. Form the product pq, and make this number public. To encode a message requires pq. To decode the message requires knowledge of the original numbers p and q. Two numbers can be multiplied together quickly, but to factorize the product is much harder. For example, 63,196,279 is the product of two prime numbers. As an exercise, find those primes!

By a large prime we mean one with over a hundred digits. Multiplying two such numbers takes a modern computer a fraction of a second. The product pq has over 200 digits, and to factorize this number into the original p and q takes years on the fastest computer. So though the product pq is public, the individual numbers p and q remain private.

Many of the topics of this book involve prime numbers. These are not just an intellectual diversion for mathematicians. The security of governments, businesses, and banks depends on prime number codes.

See: *The Enigma Machine,* pages 175–176

Fractals

Benoit Mandelbrot (1924–)

Fractals are shapes which are self-similar.

Fractals occur in both mathematics and in nature. The invention of computers has made it possible to study them.

Explaining the natural world has always been an important part of mathematics. For the most part, this has involved straight lines and smooth curves: the Earth going round the Sun in an ellipse or a body falling in a straight line. In recent years more complicated shapes have been considered.

Imagine the coastline of Britain seen from far above (Mandelbrot's own example). It looks jagged. Now imagine getting closer to the Earth. The promontories and inlets seem larger but now we can see smaller such irregularities in the coastline. The coastline still looks jagged and looks *equally* jagged as it did from further away.

This is one property of a fractal. It is self-similar. If a part of it is magnified it looks the same as it did before. An early example of a fractal was the snowflake curve. Each projection is similar to every other projection, even those much smaller or larger.

Many fractal curves have been mathematically constructed. The most famous – the Mandelbrot set – is shown. The drawing of these complicated sets would be virtually impossible without a computer.

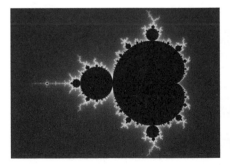

The Mandelbrot set.

The Four-Color Theorem

Kenneth Appel (1932–)
Wolfgang Haken (1928–)

Any map can be colored using only four colors.

This result was over a century old when it was finally proved. It had attracted a high number of false proofs from professional mathematicians as well as amateurs. It was proved with the aid of a computer.

In 1852 a student called Francis Guthrie was coloring a map of the counties of England. He realized that only four colors were needed, though there were many counties, as separated counties could have the same color. He wondered whether any map could be colored with only four colors and communicated this conjecture to the mathematicians of his college. It was an extraordinarily difficult problem.

There are some restrictions: two countries that meet must do so along a line – we cannot have several countries meeting at a point. A country must consist of continuous territory.

We could prove the four-color conjecture if we could produce a set of configurations with the following two properties:

1. Every map must contain one of these configurations.
2. Any map that contains one of these configurations can be colored with four colors.

Appel and Haken produced a set of configurations with these properties, in 1976. It followed that every map could be four colored. But the proof required every one of these configurations to be checked, and as they were so complicated this had to be done by computer. It took 1,200 hours of computer time and many mathematicians refused to accept a result that hadn't been checked by hand.

So far map makers have shown little interest in the result.

The Logistic Model

Robert May (1936–)

This is a model for population that takes limited resources into account.

The exponential model for population allows for indefinite growth while the logistic model introduces a factor for insufficient resources.

If a population increases at a constant percentage rate then every year it is multiplied by a constant. This is the exponential model and Malthus predicted dreadful consequences if the population was allowed to increase unchecked.

The logistic model includes a factor for limited resources. Suppose the population is P and the unchecked growth rate is r. (So, if a population is increasing at 4 per cent each year the growth rate is 0.04.) Let E stand for the equilibrium population (the population that can live in the territory without either increasing or decreasing). Then the logistic formula for the increase in population is $rP(1 - P/E)$.

Notice various things:

1. If P is very small, then the term inside the brackets is very close to one and the population grows exponentially at a rate r.
2. If $P = E$, then the term inside the brackets is zero and so there is no increase or decrease in population. The population has reached its equilibrium level.
3. If $P > E$, then the term inside the brackets is negative. So the increase is negative; the population declines, in other words. This is reasonable, as the population is greater than that which can be sustained by the territory.

The long-term behavior of the population depends on r.

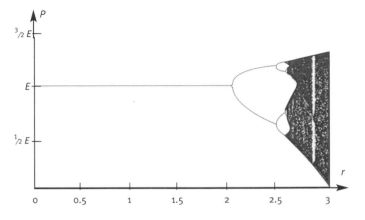

Long-term behavior of a population.

- If r is less than 2 the population settles down at the equilibrium level E.
- If r is between 2 and about 2.44, the population alternates between two fixed values, one above E and one below. We see this in nature: a year in which there is a plague of certain insects is followed by a year in which there are hardly any.
- With r between 2.45 and about 2.54, the population cycles between four fixed values.

- With r greater than about 2.6, the population varies wildly from year to year, never settling down at any value. This is an example of "chaos."

This diagram shows the long-term behavior of the population for increasing r. Notice the chaos region after about 2.6. But very strangely, a period of regularity appears in the midst of chaos, at about 2.84, when the population cycles between three fixed values. Increase r, and the system collapses back into chaos.

See: *Arithmetic and Geometric Progressions,* pages 105–106; *Chaos Theory*, page 187

Stacking Oranges

Thomas Hales (1958–)

This theorem concerns the most efficient way to stack spheres.

How should spheres be packed in order to maximize the proportion of space that they occupy? This problem was posed almost 400 years ago, yet it resisted being solved until recently.

In two dimensions, the most space-efficient way to place circles is so that each circle touches six others, as shown in the diagram.

This was proved by C. F. Gauss. The three-dimensional problem, of finding the arrangement of spheres that occupies the highest proportion of space, was posed by Johannes Kepler in 1611.

Without any help from mathematicians, greengrocers stack oranges in the following way: the bottom layer is like the arrangement in the diagram (*right*). For the next layer, put oranges in each of the hollows of the previous layer. For the third layer, place oranges in the hollows of the

second layer, and so on.

The greengrocers are right. This was shown to be the arrangement that fills the highest proportion of space. This proportion is $\frac{\pi}{\sqrt{18}}$ which is about 0.74.

Hales's proof was far from simple; indeed it was so complicated that it could not be done by hand. Like the proof of the Four Color theorem, it needed very many cases to be checked by a computer and so, as with the Four Color proof, it has been reluctantly accepted by mathematicians.

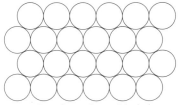

Covering a plane with circles

Fermat's Last Theorem

Andrew Wiles (1953–)

The most famous unsolved problem in mathematics is finally solved.

Fermat's last theorem states that there are no solutions in positive integers of $x^n + y^n = z^n$, if n is larger than two.

A set of positive integers for which $x^2 + y^2 = z^2$ is called a Pythagorean triple. They are the lengths of the sides of a right-angled triangle, by Pythagoras's theorem. The simplest case is $x = 3$, $y = 4$ and $z = 5$. Note that $3^2 + 4^2 = 9 + 16 = 25 = 5^2$.

Pierre de Fermat wondered whether the 2 could be replaced by a higher integer and thought that he had proved that it could not. This is Fermat's Last Theorem – that $x^n + y^n = z^n$ cannot be solved in positive integers for any n greater than 2. He teasingly wrote that he did not have enough space to write down the proof, and never communicated it to anyone.

It is very unlikely he had the full proof. Quite possibly, he had done it for the case $n = 4$ and assumed that all other cases would be similar. They are not, and progress in proving the theorem was painfully slow. The case for $n = 3$ was proved in 1770, for $n = 5$ in 1825 and for $n = 7$ in 1839. (Only prime numbers need be considered.) Then all cases up to $n = 100$ were proved but, despite many offers of prize money, no complete solution was found until 1994.

Andrew Wiles linked up the theorem with work by Japanese mathematicians on curves and produced his proof in 1993. After delivering the stunning result, he announced "*I think I'll stop here*" and sat down.

An error was found, but that was mended a year later and the proof is now agreed to be correct.

The Seven Millennium Prize Problems

Clay Mathematics Institute

These are a set of problems for the next millennium to solve.

The seven millennium prize problems are intended to be the most important outstanding problems in mathematics.

In 1900 David Hilbert produced a list of 23 problems, to summarize the progress of mathematics up to that date and to guide its progress for the century ahead. The Clay Institute in Cambridge, Massachusetts in the United States, did the same in 2000, with seven problems this time, each carrying a prize of a million dollars. The problems come from across the range of both pure and applied mathematics, with one from computing. They are as follows:

1. P versus NP.
2. The Hodge conjecture. An algebraic problem.

3. The Poincaré conjecture. This is about topology in four dimensions. It is the only one of the seven which has been solved – in 2003 by a Russian, Grigori Perelman. He refused the prize.
4. The Riemann hypothesis. This old warhorse, first stated in 1859, is the only problem to be in the lists of both 1900 and 2000.
5. Yang–Mills existence and mass gap. This concerns quantum theory.
6. Navier–Stokes existence and smoothness. This involves equations describing the flow of liquids.
7. The Birch and Swinnerton–Dyer conjecture. This is about the number of solutions of certain equations.

A major theme of this book has been the setting and solving of problems. Some were set by Greek mathematicians over 2,000 years ago, and not solved until the 19th century. Some conjectures are newer and have only recently been solved, or remain unsolved.

It is fitting that the last topic in the book should be about problems still outstanding at the start of the new millennium. Will they be solved in the next hundred years? Or will it take a thousand years? To solve these problems, what new mathematics will have to be invented?

See: *Hilbert's Problems*, page 146; *P=NP?*, pages 196–197; *The Riemann Hypothesis*, page 128

Mathematical Threads

The story of mathematics is one of a continuous development of ideas. These ideas contain threads that have run through mathematics for many thousands of years.

In many cases these threads consist of problems posed long ago and only solved much later.

The following are some threads running through the history of mathematics, which can be followed through this book.

TYPES OF NUMBER

Writing numbers
Fractions
Irrational numbers
Negative numbers
Zero
Complex numbers
Quaternions

SOLVING EQUATIONS

Quadratic equations
Cubic equations – geometric solution
Cubic equations – algebraic solution
Quartic equations

Quintic equations
Galois theory

PRIME NUMBERS

The fundamental theorem of arithmetic
The infinity of primes
A formula for prime numbers
The prime number theorem
Matiyasevich's theorem
Public key codes

STRAIGHT EDGE AND COMPASSES CONSTRUCTIONS

Regular polygons
Trisecting the angle
Doubling the cube
Squaring the circle
Regular polygons revisited
Constructible lengths
Doubling the cube and trisecting the angle revisited
Squaring the circle revisited

CALCULATING AND COMPUTING
Logarithms
Calculating machines
The difference and analytic engines
Turing machines
Binary numbers
The Enigma machine
Colossus
ENIAC
Electronic calculators
P = NP?

MODELLING THE UNIVERSE
Platonic solids
The Earth-centered universe
The Sun-centered universe
The Sun-centered universe again
Regular solids revisited
The three laws of motion
The law of gravity
Quantum mechanics
Special relativity
General relativity

WHAT IS MATHEMATICAL TRUTH?
Plato and Platonism
What the Tortoise said to Achilles
Mathematics as part of logic
Intuitionism
The Hilbert Program
Gödel's theorem

THE NATURE OF INFINITY
Zeno's paradoxes
Quadrature of the parabola
The fundamental theorem of calculus
The countability of fractions
The uncountability of the reals
The continuum hypothesis

THE DEVELOPMENT OF PROBABILITY AND STATISTICS
The problem of the points
The binomial distribution
Pascal's wager
The law of large numbers
The normal distribution
The central limit theorem

Glossary

Algebra
The manipulation of letters which represent numbers.

Algorithm
A set of rules for performing some mathematical task.

Arithmetic
The study of numbers. Basic arithmetic uses the four operations of $+$, $-$, $\div$ and x.

Axiom
An unproved statement from which other statements are deduced.

Coefficient
In an algebraic expression, the number which is multiplying a variable. In $3x^2$, the coefficient of x^2 is 3.

Conjecture
A mathematical statement that is thought to be true, but hasn't yet been proved.

Contradiction
Two statements which cannot both be true, e.g. $x > 0$ and $x < 0$.

Coordinate
The numbers which give the position of a point.

Dimension
The number of coordinates needed to give the position of a point. A line is one-dimensional, a plane is two-dimensional.

Divisor
A number that divides exactly into another number. Four is a divisor of 20, for example.

Equation
A statement which says that one expression is equal to another, such as: $x^2 = x + 3$.

Expansion
Removing the brackets in an algebraic power. The expansion of $(x + y)^2$ is $x^2 + 2xy + y^2$

Factor
This is the same as divisor.

Geometry

The study of lines, circles, spheres and so on.

Hypothesis

This is the same as conjecture.

Integer

A whole number, positive, zero or negative.

Irrational

A real number that is not rational, for example $\sqrt{2}$ or π.

Logic

The study of reasoning.

Model

A collection of mathematical expressions, equations etc., that are used to describe real life situations.

Natural number

A positive whole number. 1, 2, 3 and so on.

Paradox

An argument that leads to a contradiction.

Polygon

A closed shape in a plane with straight edges, such as a square.

Polyhedron

A solid in three-dimensional space whose faces are polygons, such as a cube.

Polynomial

An algebraic expression consisting of non-negative powers of x, multiplied by constants and added. e.g. $2x^4 - 0.5x^3 + 4x - 3$.

Power

A number, symbol or expression multiplied by itself one or more times, e.g. 3^4, x^2 or $(x + y)^5$

Prime number

A natural number that has only two divisors – one and itself, such as 2, 3, 5 and so on. Note that one is not a prime number.

Proof

An argument that will convince a mathematician of the truth of a theorem.

Qualitative
Involving considerations that do not depend on size.

Quantitative
Involving considerations that do depend on size.

Quantity
An entity that has a size, for example 5 meters or the number 5 itself.

Ratio
A measure of the relative sizes of two or more quantities, for instance the ratio of a mile to a kilometer is 1.6:1.

Rational number
An integer divided by another non-zero integer, such as 5/7, –3/1 and so on.

Real number
Any number which can be written as a decimal expansion, for example 34.387...

Set
Any collection of objects. It can be finite or infinite.

Similar
Two figures are similar if they have the same shape but not necessarily the same size. All circles are similar to each other.

Theorem
A mathematical statement that can be proved.

Index

Index

Index

Index